Air Quality and Pollution: Challenges and Concerns

Air Quality and Pollution: Challenges and Concerns

Edited by Frieda Bush

SYRAWOOD
PUBLISHING HOUSE

New York

Published by Syrawood Publishing House,
750 Third Avenue, 9th Floor,
New York, NY 10017, USA
www.syrawoodpublishinghouse.com

Air Quality and Pollution: Challenges and Concerns
Edited by Frieda Bush

© 2019 Syrawood Publishing House

International Standard Book Number: 978-1-68286-777-8 (Hardback)

Cataloging-in-Publication Data

Air quality and pollution : challenges and concerns / edited by Frieda Bush.
   p. cm.
Includes bibliographical references and index.
ISBN 978-1-68286-777-8
1. Air quality. 2. Air--Pollution. 3. Air quality management. I. Bush, Frieda.
TD883 .A37 2019
363.739 2--dc23

TABLE OF CONTENTS

PREFACE

Air pollution is caused due to the release of harmful components like greenhouse gases, biological molecules and particulates into the atmosphere. Air pollution poses the threat of critical diseases like asthma, lung cancer, cardiovascular diseases and allergies in human beings, and also causes harm to food crops and animals. Both human activities and natural processes lead to air pollution. Human activities result in the release of harmful substances into the air like carbon dioxide, nitrogen oxides, sulphur oxides, carbon monoxide, chlorofluorocarbons, etc. The main cause of air quality degradation and air pollution are anthropogenic activities, like fossil fuel burning, waste deposition in landfills, use of nuclear weapons, fumes from aerosol sprays, controlled burning practices, etc. This book provides significant information on this discipline to help develop a good understanding of air pollution and air quality degradation and their related aspects. It presents researches and studies performed by experts across the globe. This book will prove to be immensely beneficial to students and researchers in this field.

After months of intensive research and writing, this book is the end result of all who devoted their time and efforts in the initiation and progress of this book. It will surely be a source of reference in enhancing the required knowledge of the new developments in the area. During the course of developing this book, certain measures such as accuracy, authenticity and research focused analytical studies were given preference in order to produce a comprehensive book in the area of study.

This book would not have been possible without the efforts of the authors and the publisher. I extend my sincere thanks to them. Secondly, I express my gratitude to my family and well-wishers. And most importantly, I thank my students for constantly expressing their willingness and curiosity in enhancing their knowledge in the field, which encourages me to take up further research projects for the advancement of the area.

Editor

Temporal and Spatial Trends (1990–2010) of Heavy Metal Accumulation in Mosses in Slovakia

B. Maňkovská, M. V. Frontasyeva and
T. T. Ostrovnaya

Additional information is available at the end of the chapter

1. Introduction

The multielement biomonitoring surveys, using suitable plant biomonitors [1] can provide information about long-term and large-scale atmospheric deposition rates of elements. The large-scale biomonitoring programs using selected bioindicators were introduced in Slovakia in the end of 1980s. The bioindicators are commonly available elsewhere in the landscape, and the bioindicated air quality parameters can be related to the particular sampling sites within the ecosystems. mosses and foliage of forest tree species [2,3] as biomonitors of atmospheric deposition of heavy metals began in Slovakia more than 30 years ago, in connection with the problems of dying forests.

Moss species such as *Pleurozium schreberi*, *Hylocomium splendens*, and *Dicranum* sp. can effectively adsorb deposited air pollutants on pectine and cell structures. Bryomonitoring method was validated and tested to large-scale estimate current atmospheric deposition rates of elements between 1970 and 2000 [4-6]. Since 2000, in the frame of UN ECE ICP-Vegetation program, more than 30 European countries have monitored the current element content in mosses at about 7000 sampling sites in 5-year intervals. The Slovak national moss surveys since 1990–2010 have mapped elemental content distribution within the whole country (16x16 km net).

2. Material

Two complementary analytical techniques, instrumental neutron activation analysis (INAA) and atomic absorption spectrometry (AAS) were used for determination of the elemental concentrations in the samples of moss for year 2000. For INAA, moss samples of about 0.3 g were packed in aluminum cups for long-term irradiation or heat-sealed in polyethylene foil bags for short-term irradiation in the IBR-2 reactor, Dubna, described elsewhere [7]. The samples of mosses were not washed before analysis. Sulfur and nitrogen concentrations were determined using LECO corporation equipment (S: LECO SC 132 and N: LECO SC 228). Atomic absorption spectrometry (VARIAN SPECTRA A-300 and mercury analyzer AMA-254) was carried out in Forest Research Institute Zvolen (1990, 1995, 2000, 2005, 2010). The accuracy of data published in paper was verified by 109 individual laboratories and tested by the IUFRO program [8].

The monitoring studies have been undertaken in the framework of the international project Atmospheric Deposition of Heavy Metals in Slovakia Studied by the Moss Biomonitoring Technique Employing Nuclear and Related Analytical Techniques and GIS Technology. *Project REGATA (2003-2015).*

3. Results and discussion

The principal investigator of the project, Dr. Maňkovská (at that time working in the Forest Research Institute in Zvolen, Slovakia) was invited by Scandinavian specialists (Finland, UNIDO, 1986) to join the existing European biomonitoring program focused on monitoring of actual deposition of selected set of elements using analyses of mosses in 1990. The first collection of moss samples of *Pleurozium schreberi* and *Hylocomium splendens* at 58 permanent monitoring sites in Slovakia was made in the same year in accordance with the European network. The following elements were analyzed by atomic absorption spectroscopy (AAS): Cd, Cr, Cu, Fe, Mn, Ni, Pb, S, and Zn.

In the second European moss survey conducted in 1995, moss samples were collected at 78 permanent monitoring sites. In 1996, moss samples were collected at 69 and in 1997 at 74 permanent monitoring sites. The contents of As, Cd, Cr, Cu, Fe, Hg, Ni, Pb, V, and Zn were determined by AAS and Hg was determined by AMA-254.

The third moss survey at the European scale on actual levels of atmospheric deposition of elements was conducted within the ICP Vegetation in 2000. Collection of moss samples (*Pleurozium schreberi, Hylocomium splendens,* and *Dicranum* sp.) in Slovakia was performed at 86 permanent monitoring sites. NAA was carried out in the Frank Laboratory of Neutron Physics of the Joint Institute for Nuclear Research in Dubna, Russia. A total of 39 elements (Ag, Al, As, Au, Ba, Br, Ca, Ce, Cl, Co, Cr, Cs, Fe, Hf, I, In, K, La, Mg, Mn, Mo, Na, Ni, Rb, Sb, Sc, Se, Sm, Sr, Ta, Tb, Th, Ti, U, V, W, Yb, Zn, Zr) were determined. Varian Techtron atomic absorption spectrometer was used for determination of Cd, Cr, Cu, Hg, Ni, Pb, and Zn. Sulfur and nitrogen determination was performed by LECO corporation equipment as earlier (see above).

In the fourth European moss survey in 2005, moss samples (*Pleurozium schreberi, H. splendens, Dicranum* sp.) were collected at 77 permanent monitoring sites in Slovakia. They were analyzed for contents of Cd, Cu, Fe, Hg, N, Ni, Pb, S, V, and Zn by AAS Varian Techtron, AMA-2454, LECO SC 132, and LECO SP 228. Results from required monitoring elements were published in the European reports [9, 10]. Results from required and optionally monitored elements from Slovakia were evaluated in the context of neighboring countries of Visegrad Four [11].

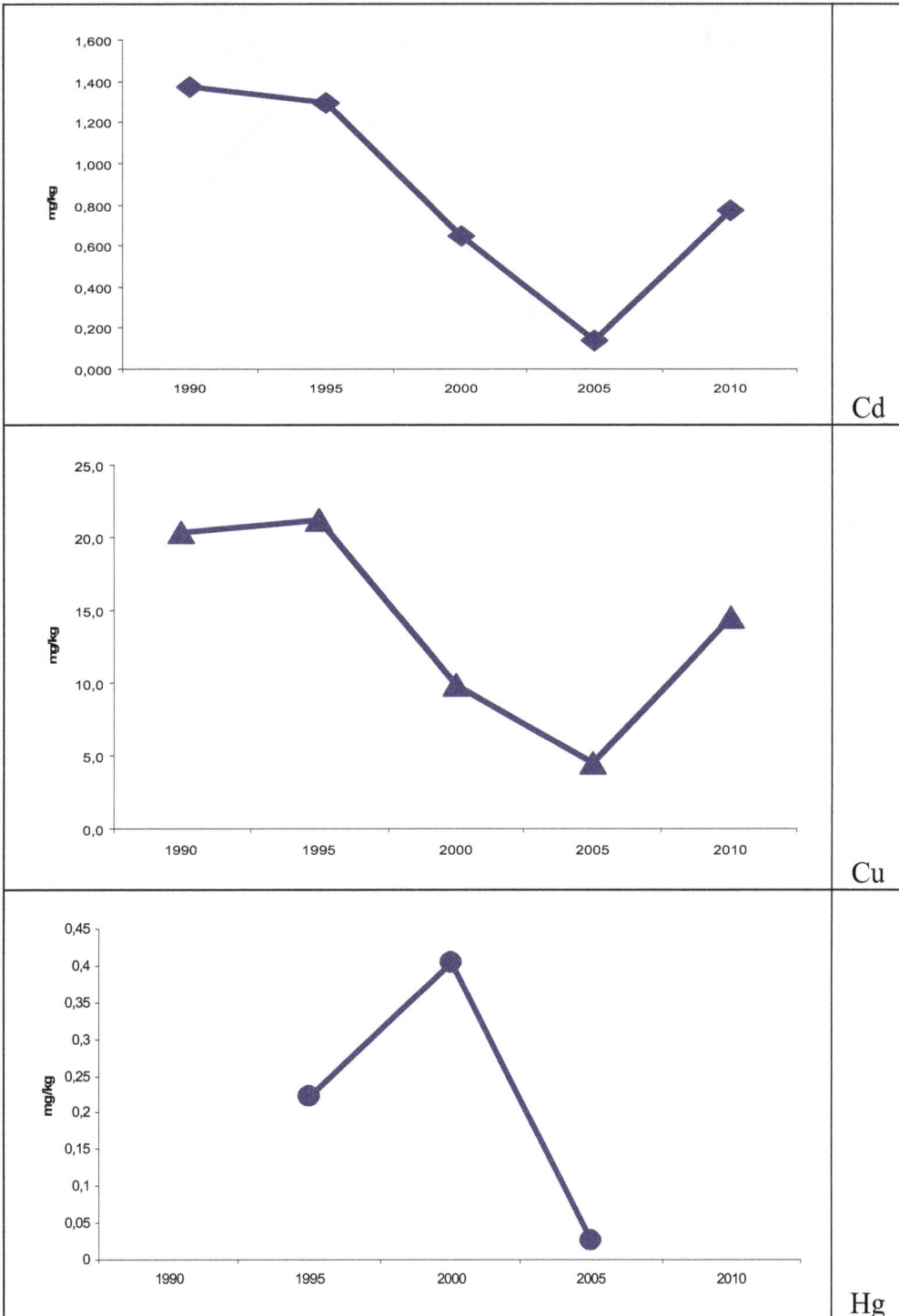

Cd

Cu

Hg

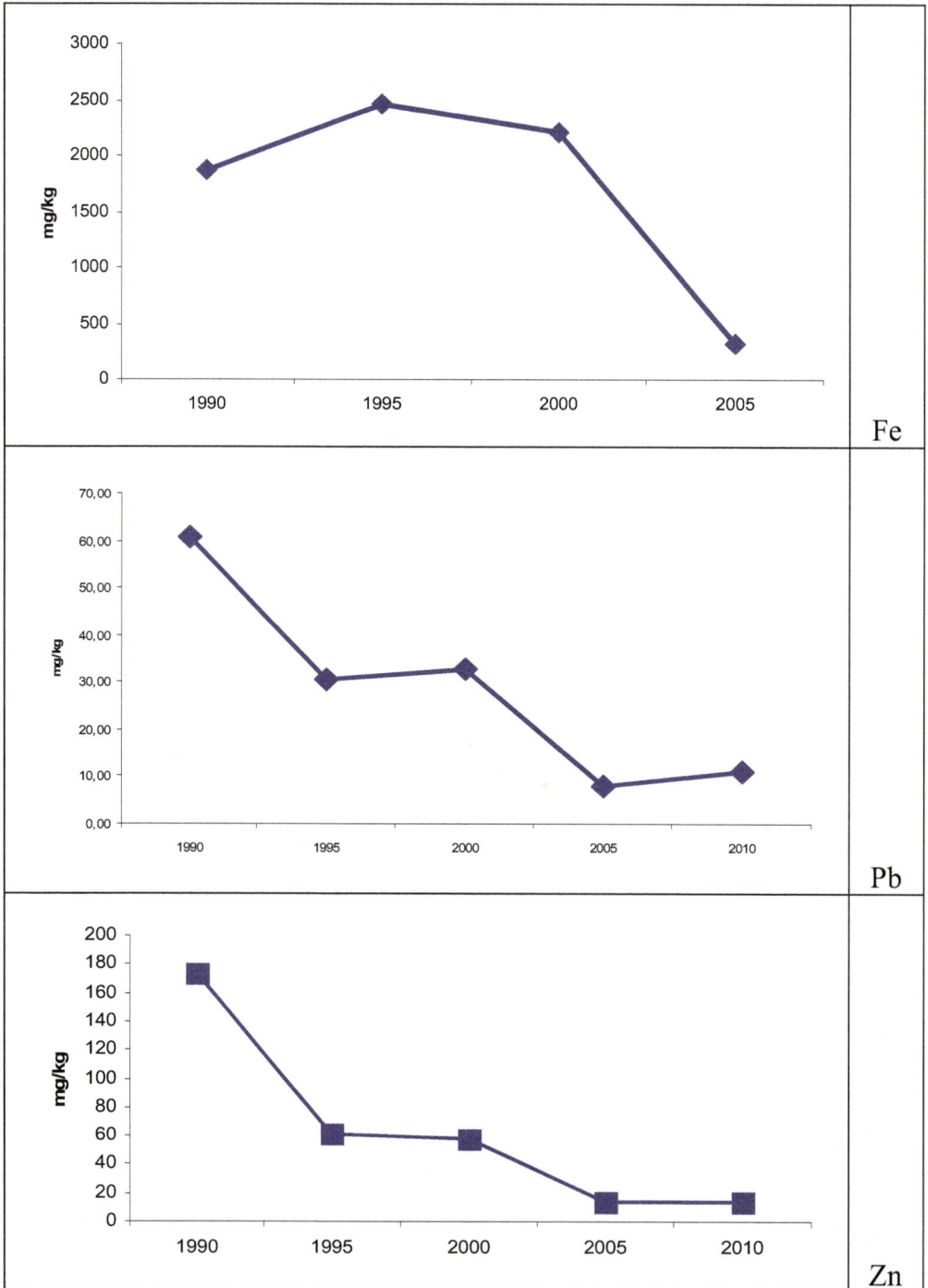

Note: Year (number of PMP): 1990(58);1995(79); 1996(69); 1997 (74); 2000 (86); 2005(82), 2010(67). PMP-permanent monitoring plots

Figure 1. Concentration of Cd, Cu, Hg, Fe, Pb, and Zn (average in mg/kg) in mosses for Slovakia in 1990, 1995, 2000, 2005, 2010.

So far, in the last, that is, the fifth European moss survey in 2010 in Slovakia, collection of moss samples was made at 68 permanent monitoring sites (*P. schreberi, H. splendens, Dicranum* sp.). They were analyzed for contents of Al, Ca, Cd, Cl, Cu, Dy, I, K, Mn, Pb, S, Ti, V by use of AAS Varian Techtron, LECO SC 132, and LECO SP 228, and by use of NAA in the Frank Laboratory of Neutron Physics in Dubna, Russia. Results from required monitoring elements were published in European reports [12].

The concentration of Cd, Cr, Cu, Fe, Hg, Ni, Pb, V, and Zn in mosses between 1990 and 2010 are shown in Fig. 1.

The moss biomonitoring technique is based on the fact that the concentration of heavy metals in mosses correlates with the atmospheric concentration. It was proven that it is possible between the concentration of the given element in mosses and the concentration of the same element in the atmosphere. The concentration of individual elements in precipitation was calculated to the time of exposure of mosses (3 years). In case of each element, there was a good linear relationship between the concentrations of a given element in mosses and in precipitation. There is a valid equation [concentration in moss] mg.kg^{-1} = [4x atmospheric deposition] mg.m^{-2}.year^{-1} [13]. The concentration of elements in mosses in comparison with Norway (Table 1 and Table 2) is expressed by means of the coefficient of loading by elements K_F and classified into 4 classes; class < 1 – elements are within norm and do not exceed the value 1; class 2 – slight loading (elements range from 1 to 10); class 3 – moderate loading (elements range from 10 to 50); class 4 – heavy loading (elements are higher than 50 times higher value).

Contamination factor K_F				
◉1	1-2	2-5	5-10	◉10
Br, I	Cl, Mn, Na, Ni, Se, Rb, U, Zn,	Ba, Ca, Co, Cr, Cu, Fe, Hg, K, Sm, Tb, Th, Ti, V	Al, Au, Ce, La, Sb, Se,Sr, Yb, Pb	Ag, Cd, Mo, Ta, W

Note: K_F = contamination factor as the rate of median values of element in Slovak mosses vs. Norvay mosses (Steinnes et al., 2007). K_F Slovakia= 9.5; K_F Norway=1.

Table 1. The rate of median values of element in Slovak vs. Norway mosses in year 2000

The marginal 2 hot spots were shown in Central Spiš (metallurgical plants), Žiar basin (nonferrous ores processing and aluminium plant). The protected area of Morské oko (chemical industry) is also of great interest. In comparison with the mean Austrian and Czech values of heavy metal contents in moss, the Slovak atmospheric deposition loads of these elements were found to be 2–3 times higher on average. The transboundary contamination by Hg through dry and wet deposition from Czech Republic and Poland is evident in the bordering territory in the north-western part of Slovakia (Black Triangle II), known for metallurgical works, coal processing, and chemical industries. Spatial trends of heavy metal concentrations in mosses were metal-specific. Since 1990, the metal concentration in mosses has declined for cadmium, chromium, cooper, iron, lead, mercury, nickel, and zinc.

Sites	Coefficient of loading by elements K_F				K_F
	< 1	1 -10	10-50	>50	
Hot Spots					
Žiar basin	Au, Br,Cl, I,Ag, In, Mn,	Al, As, Ba, Ca, Cd, Ce, Co, Cr, Cs, Cu, Fe, Hg, K, La, Mg, Mo, Na, Ni, Rb, Sc, Se, Sm, Sr, Tb, Th, Ti, U, V, W, Zn	Hf, Pb, Sb, Ta, Yb	F	6.2
Central Spiš	Au	Br, Ca, Cl, In, K, Mg, Mn, Rb, Se,	Al, As, Ba, Cd, Co, Cr, Cs, Cu, Fe, Hg, I, La, Mo, Na, Ni Sc, Sr, Th, U, V, W, Zn	Ag, Hf, Pb, Sb Ta Tb, Yb	45
National Parks					
Nízke Tatry	Au, Br, I, Mg, S, Se, Sm,Ti	Ag, Al, As, Ba, Ca, Cd, Ce, Cl, Co, Cr, Cs, Cu, Fe, Hg, In, K, La, Mn, Mo, N, Na, Ni, Pb Rb, Sb, Sc, Sr, Ta, Tb, Th, U, V, W, Yb, Zn, Zr	Hf		4.2
Vysoké Tatry	Au, Br, Ca, I, Se	Ag, As, Ba, Cd, Ce, Cl, Co, Cs, Cu, Fe, Hg, In, K, La, Mg, Mn, Mo, N, Na, Ni, Pb, Rb, S, Sc, Se, Sm, Sr, Tb, Th, Ti, U, V, W, Zn	Al, Cr, Sb, Ta, Yb, Zr	Hf	6.7
Protected Area					
Veľká Fatra	Au, Br,In Sm	Ag, Al, As, Au, Ba, Ca, Cd,Ce, Cl, Co, Cs, Cu, Fe, Hg, I, K, La, Mg, Mn, Mo, N,Na, Ni, Pb Rb, S, Sc, Se, Sr, Ti, U, V, W, Zn	Cr, Sb, Ta, Tb, Th, Yb, Zr	Hf	7.6
Báb	Au, Br, In, Mg, N, S, Se	Ag, As, Ba,Ca, Cl, Co, Cr, Cs,Cu, Fe,Hg, I, K, Mn, Na, Ni, Rb, Sm, Sr, Ti, U, V, W, Zn	Al, Cd, Ce, La, Mo, Pb, Sb, Sc, Ta, Tb, Th,Yb, Zr	Hf	8.8
Slovenský raj	Au, Br, In Sm, Se	Al, As, Ba, Ca, Cd, Ce, Cl, Co, Cr, Cs, Cu, Fe, I, K, La, Mg, Mn, N, Na, Ni, Rb, S, Sc, Sr, Th, Ti, U, V, W, Zn	Ag, Hg, Mo, Pb, Ta, Tb, Yb, Zr	Hf, Sb	11.8
Poľana	Au	Br, Ca, Cl, Cu, In, K, Mg, Mn, Na, Rb, Se, Zn	Ag, Al, As, Ba, Cd, Co, Cr, Cs, Fe, Hg, I, La, Mo, Ni, Pb, Rb, Sc, Sr, Ta, Tb, Th, U, V, W, Yb	Sb, Hf	19
Morské oko	Au	Br,Ca, Cl, In, K, Mg, Mn, Rb, Se, Zn	Ag, As, Ba, Cd, Co, Cr, Cs, Cu, Fe, Hg, I, La, Mo, Na, Ni, Pb, Sr, Sc, Ta, Tb, U, V, W	Al, Hf, Sb, Th, Yb	44

Table 2. Coefficient of loading by elements K_F in the year 2000

The temporal trends in the concentration of Cd, Cr, Cu, Fe, Hg, Ni, Pb, V, and Zn between 1990 and 2010 were observed. In general, the concentration of Cd, Cr, Cu, Fe, Hg, Ni, Pb, V, and Zn in mosses decreased between 1990 and 2010; the decline was higher for Pb than for Cd. The observed temporal trends for the concentrations in mosses were similar to the trends reported for the modeled total deposition of cadmium, lead, and mercury in Europe. The level of elements determined in bryophytes reflects the relative atmospheric deposition loads of the elements at the investigated sites. Factor analysis was applied to determine possible sources of trace element deposition in the Slovakian moss. In the industrial area of Central Spiš, in comparison with the Norwegian limit values (Central Norway is considered a relatively pristine region), exceeded levels for Al, As, Ca, Cd, Cl, Co, Fe, K, Mn, Sb, Sm, Sr, W, and Zn were found.

4. Conclusion

Moss surveys can provide quick and cheap information about spatiotemporal changes of the current deposition rates of about 40 chemical elements across the country. Figures from the moss surveys may be the only data about elemental deposition rates that have not been determined at measurement stations of air quality (e.g., Be, Li, Se, Tl, Th, and REEs).

Moss biomonitoring is an effective tool for detecting effects of new technologies on deposition zones in the vicinity of emission sources. All results of the Slovak moss surveys were accepted and stored in the UN ECE ICP-Vegetation database for checking of deposition loads in Europe and their environmental effects.

On the basis of biomonitoring using 3-year-old segments of *Pleurozium schreberi*, *Hylocomium splendens*, and *Dicranum* sp. at 10 sites in Slovakia, it was determined that:

a. The concentration of elements (in parentheses) is more than 50 times higher at sites Báb (Hf), Poľana (Hf, Sb); Vysoké Tatry (Hf); Slovenský raj (Hf, Sb); Veľká Fatra (Hf); Central Spiš (Ag, Hf, Pb, Sb Ta Tb, Yb); Žiar basin (F), and site Morské oko (Al, Hf, Sc, Sb, Ta, Tb, Th, Yb) compared to the Norwegian values.

b. Air pollutants K_F varies in the range of 4–45 (4.2 – Nízke Tatry; 6.2 – Žiar basin; 6.7 – Vysoké Tatry; 7.6 – Veľká Fatra; Báb – 8.8; 11.8 – Slovenský raj; 19 – Poľana; 44 – Morské oko; and 45 – Central Spiš). Results of biomonitoring campaigns serve as a reliable basis for planning and long-term exploitation of the landscape of the country and for further environmental investigations.

Acknowledgements

This article was made possible with the financial support of grant APVV-0663-10, VEGA and by the grant of the Plenipotentiary of the Slovak Republic at the Joint Institute for Nuclear Research, Dubna, Russian Federation.

Author details

B. Maňkovská[1*], M. V. Frontasyeva[2] and T. T. Ostrovnaya[2]

*Address all correspondence to: bmankov@stonline.sk

1 Institute of Landscape Ecology, Slovak Academy of Sciences, Bratislava, Slovakia

2 Frank Laboratory of Neutron Physics, JINR, Dubna, Russian Federation

References

[1] Markert, B.A., Breure, A.M., Zechmeister, H.G., 2004: *Bioindicators and Biomonitors. Principles, Concepts and Applications*. 2nd Edition, 997 pp. Amsterdam – Tokyo: Elsevier.

[2] Maňkovská, B., 1995: Mapping of forest environment load by selected elements through the leaf analyses. *Ecology* (Bratislava) 14, 2, 205-213.

[3] Maňkovská, B., 1996: Geochemický atlas Slovenska - Lesná biomasa. Geologická služba Bratislava, ISBN 80-85314-51-7, 87 pp.

[4] Rühling, A., Tyler, G., 1970: Sorption and retention of heavy metals in the woodland moss *Hylocomium splendens* (Hedw.) *Br et Sch Oikos*, 21(1), 92–97.

[5] Ross, H.B., 1990: On the use of mosses (*Hylocomium splendens* and *Pleurozium schreberi*) for estimating atmospheric trace metal deposition. *Water Air Soil Poll*, 50(1-2), 63–76.

[6] Berg, T., Steinnes, E., 1997: The use of mosses (*Hylocomium splendens* and *Pleurozium schreberi*) as biomonitors of heavy metal deposition from: From relative to absolute deposition values. *Environ Poll*, 98(1), 61–71.

[7] Frontasyeva, M.V., 2011: Neutron activation analysis for the Life Sciences. A review: "Physics of Particles and Nuclei", 42, 2, p. 332-378 (in English). http://www.springer-link.com/content/f836723234434m27

[8] Maňkovská, B., Oszlányi, J., 2010: Concentration of 45 elements in moss and their temporal and spatial trends in Slovakia (1990-2005). In *Landscape Ecology - Methods, Applications and Interdisciplinary Approach*. Bratislava. Institute of Landscape Ecology Slovak Academy of Sciences, p. 341-351. ISBN 978-80-89325-16-0.

[9] Schröder, W., Pesch, R., Englert, C., Harmens, H., Suchara, I., Zechmeister, H.G., Thöni, L., Maňkovská, B., Jeran, Z., Grodzinska, K., Alber, R., 2008: Metal accumulation in mosses across national boundaries: Uncovering and ranking causes of spatial variation. *Environ Pollut*, 151: 377–388.

[10] Harmens, H., Frontasyeva, M., Maňkovská, B., et al., 2010: Mosses as biomonitors of atmospheric heavy metal deposition: spatial and temporal trends in Europe. *Environ Poll*, 158: 3144-3156.

[11] Suchara I., Florek M., Godzik B., Maňkovská B., Rabnecz G., Sucharova J., Tuba Z., Kapusta P., 2007: Mapping of Main Sources of Pollutants and their Transport in Visegrad Space. Silvia Taroucy Institute for Landscape and Ornamental Gardening Průhonice, CZ, ISBN 978-80-85116-55-7.

[12] Harmens, H., Ilyin, I., Mills, G., Aboal, J.R., Alber, R., Blum, O., Coskun, M., De Temmerman, L., Fernandés, J.Á., Figueira, R., Frontasyeva, M.V., Godzik, B., Goltsova, N., Jeran, Z., Korzekwa, S., Kubin, E., Kvietkus, K., Leblond, S., Liiv, S., Magnússon, S.H., Maňkovská, B., Nikodemus, S., Pesch, R., Pikolainen, J., Radnovič, D., Rühling, A., Santamaria, J.M., Schröder, W., Spiric, Z., Stafilov, T., Steinnes, E., Suchara, I., Tabors, G., Thöni, L., Turcsányi, G., Yurukova, L., Zechmeister, H.G., 2012: Country-specific correlations across Europe between modelled atmospheric cadmium and lead deposition and concentrations in mosses. *Environ Poll*, 166, 1–9.

[13] Steinnes, E., Berg, T., Uggerud, H., Vadset, M., 2007: Atmospheric deposition of heavy metals in Norway (in Norwegian). Nation-wide survey in 2005. State Program for Pollution Monitoring, Report 980/2007. Norwegian State Pollution Control Authority, Oslo 2007, 36 pp.

## 2

Ambient Level of NO$_x$ and NO$_y$ as Indicators of Photochemical Activity in an Urban Center

Alberto Mendoza and Edson R. Carrillo

Additional information is available at the end of the chapter

1. Introduction

Atmospheric pollution is considered a severe problem, especially in large urban areas where anthropogenic emissions (e.g., emissions from domestic, industrial, and transportation activities, as well as from other productive sectors) mix with biogenic emissions (i.e., emissions with natural origins). Anthropogenic emissions include gas-phase primary air pollutants such as nitric oxide (NO), nitrogen dioxide (NO$_2$), carbon monoxide (CO), volatile organic compounds (VOCs), and sulfur dioxide (SO$_2$). Although these pollutants can produce harmful health effects, their ability to react as precursors of secondary air pollutants is one of their most relevant characteristics.

Ozone (O$_3$) is produced as a secondary pollutant by photochemical reactions occurring between nitrogen oxides (NO$_x$, where NO$_x$ = NO + NO$_2$) and VOCs in the presence of sunlight [1, 2]. For a long time, interest in O$_3$ focused on its direct health effects as a major constituent of photochemical air pollution [3, 4] and its impacts on vegetation [5, 6]. However, O$_3$ also affects the energy budget of the atmosphere; thus, it has become part of a family of species referred to as short-lived climate pollutants (SLCP) [7].

Photochemical activity occurs in both natural and human-altered environments. Theoretically, in the presence of only NO$_x$ in the troposphere, O$_3$ generation could be described by a simple mechanism known as the NO$_x$-O$_3$ photostationary state, summarized by the following reactions:

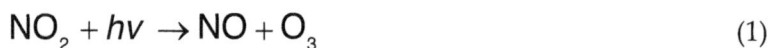

$$NO_2 + hv \rightarrow NO + O_3 \tag{1}$$

$$O^\bullet + O_2 + M \rightarrow O_3 + M \tag{2}$$

$$O_3 + NO \rightarrow NO_2 + O_2, \tag{3}$$

where hv represents a photon (sunlight energy), $O^\bullet$ denotes an oxygen free radical (an oxygen atom with an unpaired electron), and M is known as a third body (molecule) that acts as an energy sink. In the atmosphere, this third body is typically N_2 or O_2. Under this chemical reaction scheme, the net O_3 production is zero, and according to the photostationary state equation, the concentration of O_3 can be determined based on the concentration of NO_x and the amount of solar radiation:

$$\left[O_3\right]_{ps} = \left(\frac{j_{NO_2}}{k_{R3}}\right)\frac{\left[NO_2\right]}{\left[NO\right]}, \tag{4}$$

where k_{R3} is the kinetic reaction rate constant for reaction (3) and j_{NO2} is the photolysis rate of NO_2. As a demonstration, if a 10 parts per billion (ppb) value is considered for the term j_{NO2}/k_{R3}, the O_3 concentration would be 27 ppb, with an initial value of 100 ppb of NO [8]. An O_3 concentration ranging between 10–40 ppb is typical of rural areas around the globe [9].

When VOCs are added to the mixture of species present in the troposphere, the observed O_3 levels are higher than those predicted by the photostationary state formulation. In this case, interactions in the O_3-NO_x-VOC system are initialized by the hydroxyl ($HO^\bullet$) radical, through the photolysis of O_3:

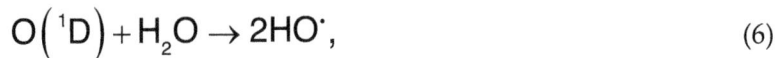

$$O_3 + hv \rightarrow O_2 + O\left(^1D\right) \tag{5}$$

$$O\left(^1D\right) + H_2O \rightarrow 2HO^\bullet, \tag{6}$$

where $O(^1D)$ represents an excited singlet oxygen atom. The oxidation of the VOCs (represented here as RH, where R is an organic functional group, e.g., an alkyl group) continues in the presence of $HO^\bullet$ and NO_x to produce more O_3 and a variety of nitrogen-containing species. The process can be represented by the following generalized reactions [8]:

$$RH + HO^\bullet + O_2 \rightarrow RO_2^\bullet + H_2O. \tag{7}$$

The alkyl peroxide radical ($RO_2^\bullet$) reacts with NO to form aldehydes (R'CHO) and peroxide radicals ($HO_2^\bullet$):

$$RO_2^{\cdot} + NO + O_2 \rightarrow NO_2 + R'CHO + HO_2^{\cdot} \qquad (8)$$

Reaction 8 is a relevant process because it represents an alternate route for the production of NO_2 from NO, but without destroying O_3 (in contrast to reaction [3]). A similar process occurs when the aldehydes continue reacting to eventually form acyl peroxy radicals ($R'C(O)O_2^{\cdot}$) that undergo a similar fate as the $RO_2^{\cdot}$ radicals in reaction (8). Thus, O_3 starts to accumulate in the system, as it is no longer destroyed by reaction (3). It continues to be produced by reactions (1) and (2) as NO_2 and NO cycle through this set of reactions. In addition, the $RO_2^{\cdot}$ radicals can produce nitrates when they react with NO_2:

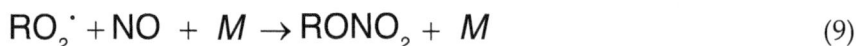

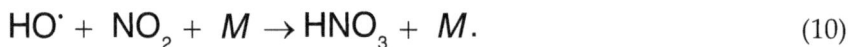

$$RO_2^{\cdot} + NO + M \rightarrow RONO_2 + M \qquad (9)$$

$$HO^{\cdot} + NO_2 + M \rightarrow HNO_3 + M. \qquad (10)$$

This description is not exhaustive, as we have focused our attention on the main reactions of O_3 production by NO_x and VOCs. As indicated, NO_x acts as catalyst in these reactions, while the VOCs continue to undergo oxidation until they are converted to CO_2. In parallel, a variety of different inorganic (e.g., HNO_3) and organic (e.g., $RONO_2$) nitrogen-containing species are also produced. Some of these substances act as reservoirs of NO_x which are released to the reacting mixture upon decomposition (e.g., peroxy acetyl nitrate or PAN, an organic nitrate), and others act as sinks (most notably, HNO_3). The sum of NO_x and these additional inorganic and organic nitrates is referred to as total reactive nitrogen or total odd nitrogen oxides (NO_y).

In general, the O_3-NO_x-VOC system increases the NO_x available to react through reactions (1)-(2) (increasing O_3), thereby producing a complex mixture of partially oxidized VOCs that mixes with freshly emitted VOCs and different oxidized nitrogen species (HNO_3, HNO_2, NO_3, PAN, etc.). The study of the dynamics of O_3 production can be quite challenging, given the complexity of the chemical mixtures and their nonlinear response to emission changes and meteorological conditions. For this reason, comprehensive air quality models have been devised to study the complex physical and chemical processes that participate in gas-phase production of O_3 and other air pollutants [10].

Observational-based approaches have proven to be valuable for describing the conditions and regimes that foster air pollution. Some observational approaches use considerable amounts of data that can be obtained through networks of routine air quality and meteorological monitoring stations, and conclusions are then inferred based on the statistical analysis of these data [11-17]. The analyses of these databases can be enhanced if information is also available for additional indicator species that are typically not routinely monitored, such as NO_y.

In this study, we have analyzed data gathered by the routine air quality monitoring stations of the Monterrey Metropolitan Area (MMA; 25° 40′ N, 100° 18′ W). Monterrey is the third most populated urban center in Mexico (4.1 million inhabitants), second in size in the country in terms of industrial infrastructure, and one of the cities with the worst air quality problems in Mexico [18]. The MMA (see Figure 1) has been in violation of the 1-hr Mexican Air Quality Standard for O_3 (0.11 ppm) since the establishment of its routine air quality monitoring system in 1993. For example, the 1-hr O_3 standard was exceeded on 48 different days in 2011 [19]. Peak O_3 concentrations can reach 170 ppb and typically occur at the downtown or western air quality stations [20]. Some studies report that the MMA is the fifth most polluted Latin American city in terms of O_3 [21]. In addition to the large number of industrial facilities located inside or nearby the metropolitan area that contribute to the poor air quality, emissions from several gas-fired electric utilities and one of the six refineries that operate in the country (located less than 40 km to the east of downtown Monterrey) add to the anthropogenic burden imposed on the airshed. Despite the importance of the contribution of area and point sources to the total emissions inventory, the mobile sources represent approximately 75% of the total anthropogenic emissions released in the MMA [20]. This is a result of a relatively high proportion of vehicles per inhabitant registered in this urban center (approximately one vehicle for every two inhabitants).

2. Methods: Database and analysis tools

The study used data collected by the *Sistema Integral de Monitoreo Ambiental* (SIMA; Integrated Environmental Monitoring System) of the MMA. At the time of the study, valid data from six operational routine air quality stations were available: Downtown, Southeast, Southwest, Northeast, Northwest, and North (see Figure 1). Data archived for the period of August 2012 to August 2013 were retrieved. The database contained hourly-average concentrations of CO, NO, NO_2, and O_3, as well as meteorological parameters (relative humidity, atmospheric pressure, dry bulb temperature, solar radiation [SR], wind speed [WS], and wind direction [WD]). These data go through a quality assurance/quality control process defined by SIMA in compliance with international standards. Additional air quality parameters monitored by the stations (mainly, SO_2 and particulate matter with aerodynamic diameter less than or equal to 10 microns [PM_{10}] and less than or equal to 2.5 microns [$PM_{2.5}$]) were not used in this study, as the focus was on the relationship between O_3 and NO_x.

In addition to the above chemical parameters, NO_y was monitored in the Downtown station from August 2012 to August 2013, using a NO-NO_y (Thermo Scientific, Model 42i-Y) chemiluminescence continuous sampling device. NO_y is a chemical parameter that is not routinely measured by Mexican air quality stations, as it is not considered a criteria pollutant. The decision to deploy the NO-NO_y instrument at the location of the Downtown station was based on spatial homogeneity studies performed in the past for the MMA [22], which indicate that pollutant levels observed in the Downtown station are representative of the MMA, with the exception of the Southeast region. The data collected by the NO-NO_y device went through a validation process similar to that routinely conducted by SIMA on its own collected data. The main conditions to reject data included the following red flags: obstructed capillary tubing, low flow in the inlet, and concentrations outside the measurement range (i.e., above 200 ppbv). The data were then consolidated and analyzed on a daily basis, a monthly basis, and by seasons (Summer and Fall 2012, and Winter, Spring, and Summer 2013). Summer 2012 included only data from August and September of 2012; Summer 2013 did not include data for September 2013.

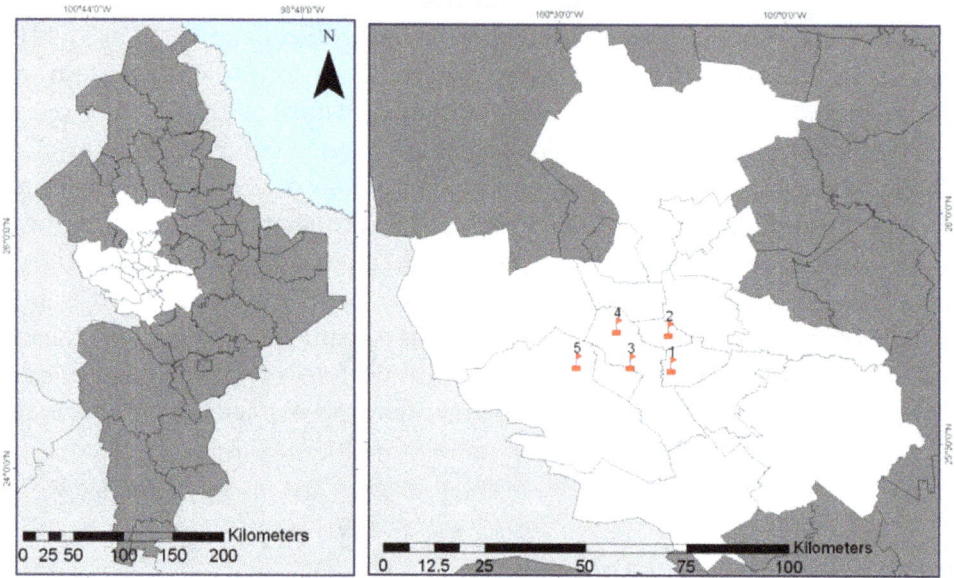

Figure 1. Municipalities that comprise the Monterrey Metropolitan Area (left panel) and location of the air quality stations monitoring stations used in this study (right panel): 1 Southeast, 2 Northeast, 3 Downtown, 4 Northwest, 5 Southwest; data from the North station was not used

Once the data were consolidated, an exploratory analysis of the data set was conducted to derive descriptive statistics. In addition, correlation analysis was conducted to explore relations between the chemical and meteorological parameters, with emphasis on the relationships among O_3, NO_x, and NO_y. This analysis was complemented by the use of polar plots and wind roses to determine the relationship between high pollutant levels and transport conditions (wind speed and direction).

Grouping techniques were then applied for further exploration of the data. Two methods were used: Principal Components Analysis (PCA) and Analysis of Variance (ANOVA). PCA is helpful in reducing the dimensionality of the dataset, and ANOVA can identify differences among data groups. Thus, PCA was used to identify the most important variables in the dataset which in turn merit further exploration. With the ANOVA, we attempted to determine differences in the weekday and weekend conditions that resulted in high O_3 levels. All statistical analyses were conducted using Minitab® 16, and plots were constructed using the R programming language.

3. Results

3.1. Descriptive statistics

Table 1 summarizes the descriptive statistics of the chemical and meteorological parameters retrieved from the Downtown air quality monitoring station. The values presented are based on the daily means for the August 2012-August 2013 period. The climatology of the MMA is portrayed in these results: a hot semi-arid region characterized by extreme weather conditions [23]. The long-term average temperature of Monterrey is around 23°C, with temperatures that can reach 40°C in summer and below 0°C in winter. Average humidity is 62%; the rainy season is between August and October, directly related to the occurrences of hurricanes in the Gulf of Mexico (the average mean annual precipitation is around 600 mm). Due to its geographic location, the MMA is influenced by anti-cyclonic systems from the Gulf of Mexico [24], which in some instances result in high atmospheric stability in the region, thereby inhibiting the vertical mixing of pollutants.

Species/Parameters	Mean ± Std. Dev.	Median	Maximum	Minimum
O_3 (ppb)	25.7 ± 9.0	24.4	50.7	2.6
NO_2 (ppb)	16.3 ± 5.9	15.3	39.2	3.3
NO (ppb)	6.5 ± 4.7	5.3	34.4	0.6
NO_y (ppb)	40.0 ± 20.3	35.2	188.2	12.1
CO (ppm)	0.78 ± 0.26	0.75	1.78	0.21
Solar radiation (kW/m²)	0.24 ± 0.12	0.24	0.70	0.00
Temperature (°C)	24.4 ± 5.2	25.8	32.5	9.5
Wind speed (km/h)	1.84 ± 0.47	1.84	3.21	0.71
Relative Humidity (%)	62.9 ± 16.8	64.17	98.9	13.5

Table 1. Descriptive statistics for parameters monitored at the Downtown air quality station of the MMA (annual means, medians, maximum values, and minimum values of the daily averages)

Details of the monthly variation of the daily averages of the chemical and meteorological parameters are depicted in Figure 2. O_3 levels tend to be the highest during springtime, as well as during late summer and early fall. These periods correspond to the times when temperature and SR are high and wet precipitation is low, i.e., conditions that foster photochemical reactions. Another relevant feature presented in Figure 2 involves the inverse relationship observed between O_3 and NO_x. That is, during the cold months, when SR is low, NO_x levels tend to be high due to inhibition of the reactions between VOCs and NO_x that promote the conversion of NO to NO_2, as well as suppression of the photo-dissociation of NO_2 to produce O_3. We need to bear in mind that NO_x emissions are mainly in the form of NO. Thus, an accumulation of NO and NO_2 is observed in winter. At other times, temperatures and SR increase, promoting photochemical reactions. A meteorological component influences the observed NO_x concentrations: shallow mixing heights and low wind speeds during winter. The O_3 seasonal boxplots indicate that even though fall is the second most important season with respect to the frequency of peak O_3 levels, it is also the time when the average O_3 is the lowest (Figure 3). This can be explained by the incidence of rain events associated with this season.

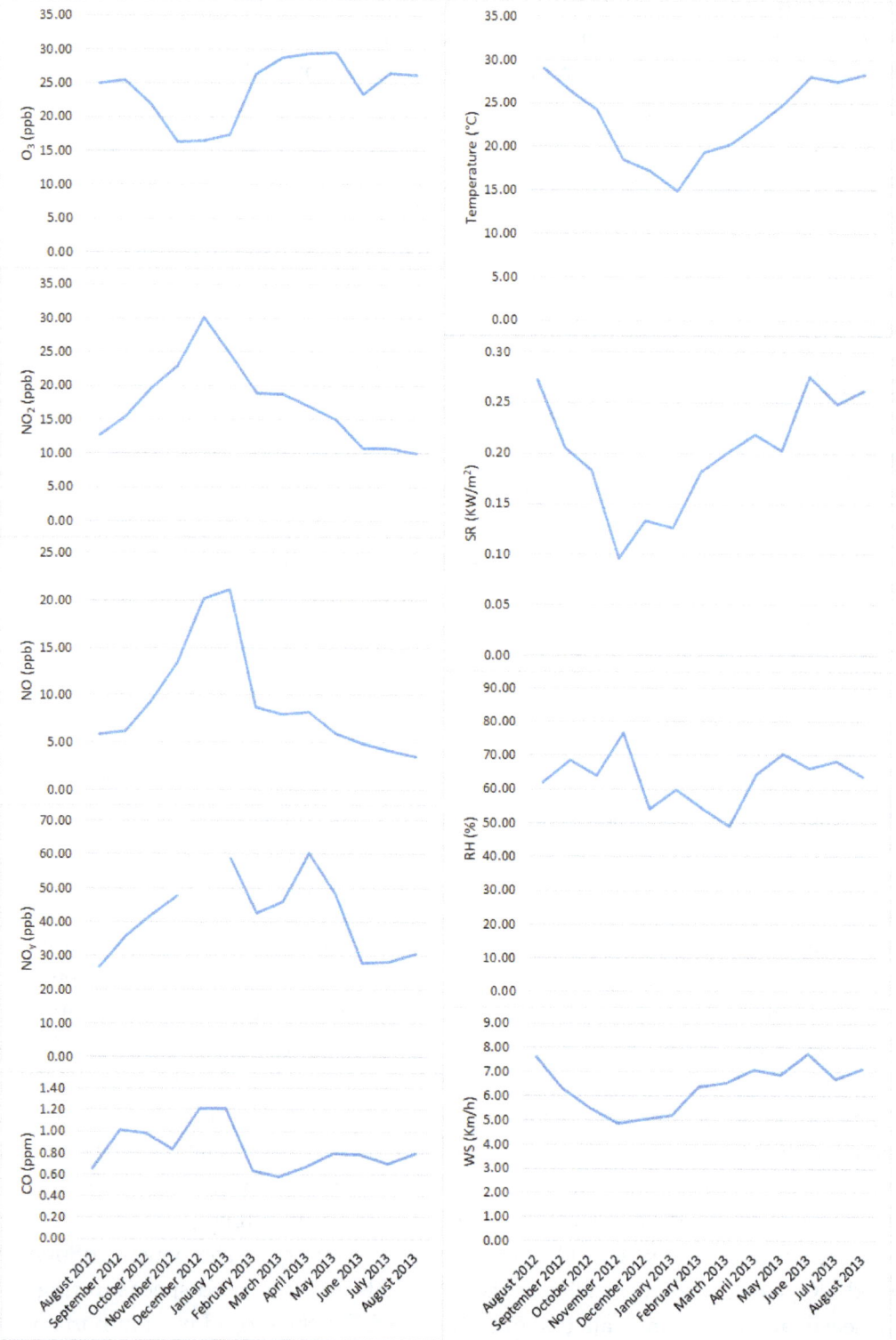

Figure 2. Time series plots of monthly averages of the mean daily values of chemical and meteorological parameters observed at the Downtown station

3.2. Spatial analysis

The exploratory analysis was complemented by the construction of bivariate polar plots of NO_x (see Figure 4) and wind roses (see Figure 5) for five of the six air quality monitoring stations, covering the main sub-regions of the MMA. The five selected stations correspond to typical upwind locations (Northeast and Southeast), Downtown Monterrey, and typical downwind locations (Northwest and Southwest), as illustrated in Figure 5. Seasonal variation is also presented in these plots.

The polar plots represent NO_x concentration as a function of wind speed and direction, and they help identify the possible occurrence/prevalence of horizontal transport conditions that lead to high NO_x events. Further away from the center of the plot, the wind speed is higher. The form of the polar plot can provide an indication of the wind direction and speed frequency; however, this information is better observed through the corresponding wind roses.

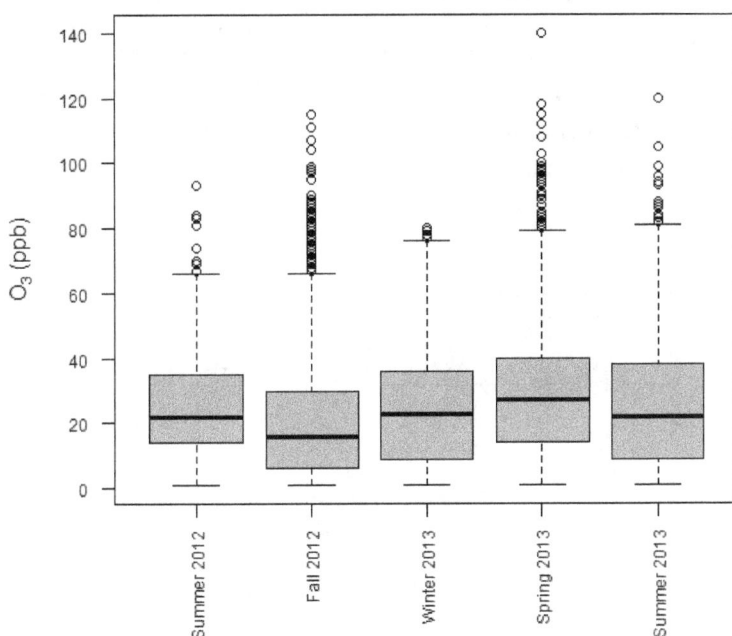

Figure 3. O_3 boxplots by seasonal periods for 1-hr average values reported at the Downtown station

In general, Figure 4 indicates higher transport of NO_x from the northern stations (this can be catalogued as "fresh" NO_x), which then is consumed as the air masses traverse the urban core. The transport patterns appear to be similar in all seasons. Thus, the change in NO_x by season appears to be more strongly linked to photochemical conditions (temperature and SR) than to wind patterns.

Using the same analysis as was used for NO_x, Figure 6 depicts the behavior of O_3 as a function of wind speed and direction exclusively at the Downtown station. In all cases, the higher concentrations begin to occur when the wind velocity was above 2 m/s, indicating the influence of transport and aging of air masses on the observed O_3 concentrations and weak O_3 production under stagnant conditions. Transport influences the levels of O_3; consequently, the highest O_3 levels correlate with the prevailing wind directions (wind blowing from the Northeast-East-Southeast and from the North-Northwest-West). The East-to-West air flow channels air masses through the long axis of the MMA, adding more precursors to the chemically aged mixtures as they traverse the urban core, thereby promoting O_3 production.

The frequency of high O_3 concentrations related to a specific wind direction cannot be inferred from the polar plots, however. That analysis was conducted by constructing boxplots for 1-hr average O_3 concentrations for the three main wind categories (classified based on direction from which the wind was blowing) and segregated by season (see Figure 7). The year around the peak O_3 concentrations and the maximum 1-hr average concentrations occur when the wind blows from the Northeast-East-Southeast. O_3 levels, on average, are lower when the wind blows from the North-Northwest-West, and they are the lowest when winds blow from the South-Southeast (with the exception of the summer, when this direction provides higher average concentrations than the North-Northwest-West direction).

Comparison of the O_3 levels presented by the polar plots (Figure 6) for spring and fall indicates that the influence of transport is less important during fall than in spring, as wind speeds tend to be lower; high O_3 events are mainly related to events when the speed is in the range of 2.0-3.5 m/s. In spring, O_3 levels can remain high at wind speeds above 4 m/s.

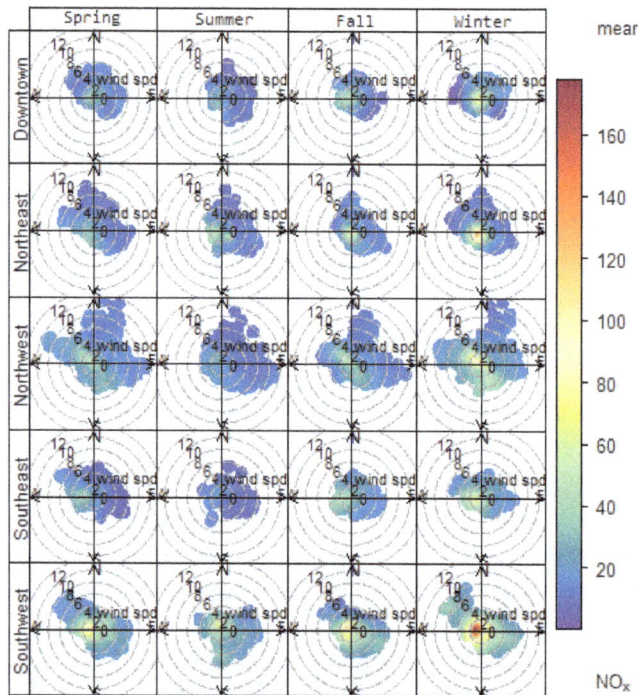

Figure 4. Polar plot for 1-h average NO_x concentrations at five different monitoring stations within the MMA (the radial dimension is an indication of increasing wind speed [m/s])

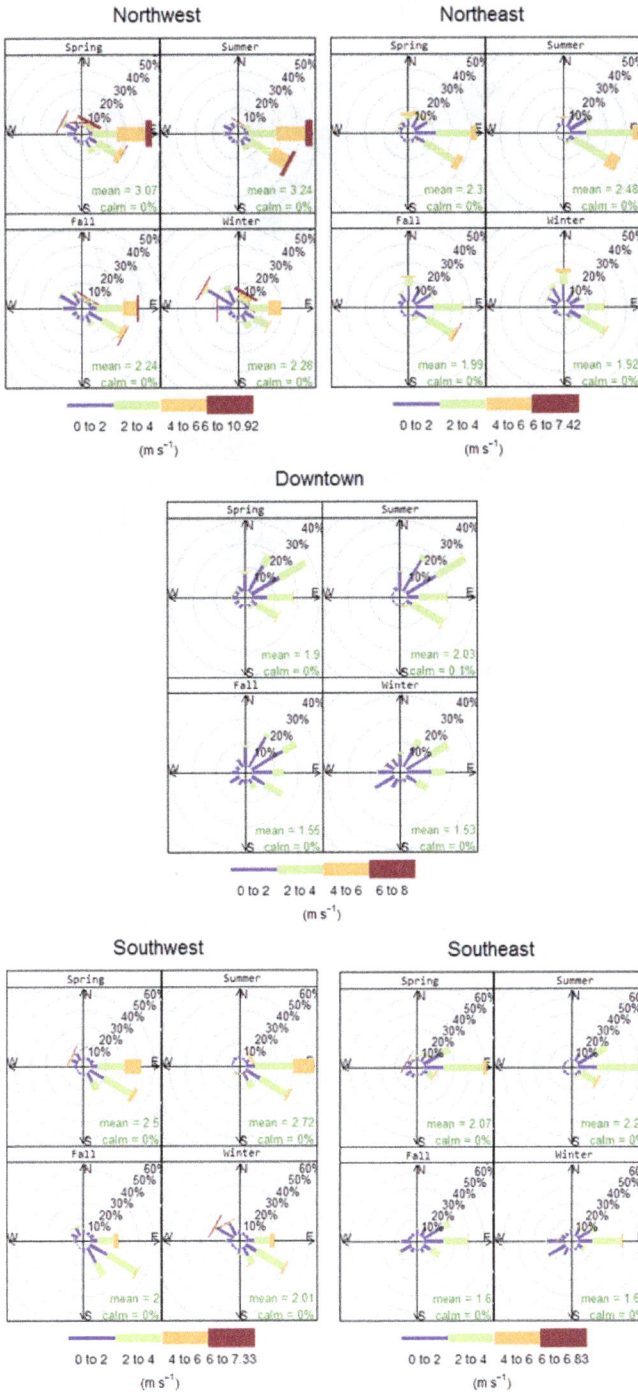

Figure 5. Wind roses at five different monitoring stations within the MMA

Downtown O_3

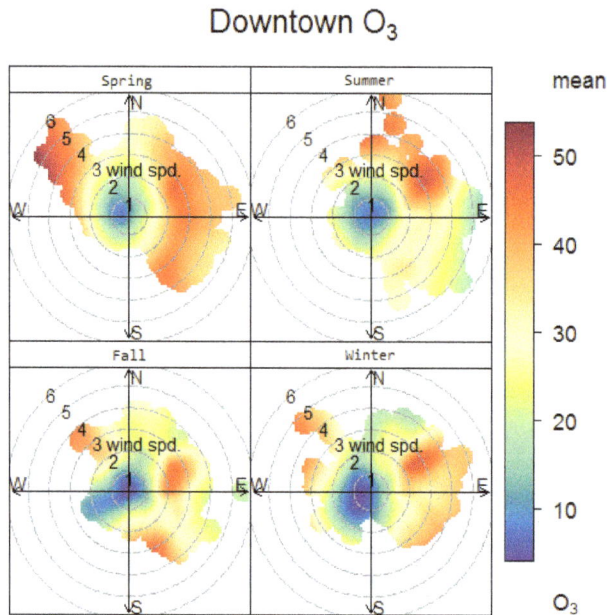

Figure 6. Polar plot for 1-h average O_3 concentrations at five different monitoring stations within the MMA (the radial dimension is an indication of increasing wind speed [m/s])

3.3. Regression analysis

Figure 2 gives an initial indication of a direct relationship between NO and NO_2 levels and an inverse relationship between NO_x and O_3 levels. This was explored further by means of scatter plots (not presented for brevity), which were used to derive linear regression expressions among NO_x constituents and between NO_x and O_3:

$$\left[NO_x \right] = 1.5 \left[NO \right] + 13.0 \tag{11}$$

$$\left[O_3 \right] = -0.37 \left[NO_x \right] + 34 \tag{12}$$

where [i] represents the molar concentration of species i. Here, the 1-hr average concentrations were used to derive equations (11) and (12). These expressions corroborate prior findings. The relevance of equation (12) is that it provides a measure of the expected, first order change in O_3 provided by a change in NO_x concentration (which can occur from a change in NO emissions). The results indicate that O_3 would, in fact, increase as NO_x tends to be reduced. This counterintuitive finding is a well-known result seen in urban areas dominated by a VOC-sensitive regime [25]. Other studies that have used complex photochemical air quality models have reached similar conclusions. Downtown Monterrey appears to be dominated by a VOC-sensitive regime in which VOC controls would work better for reducing O_3 levels [20].

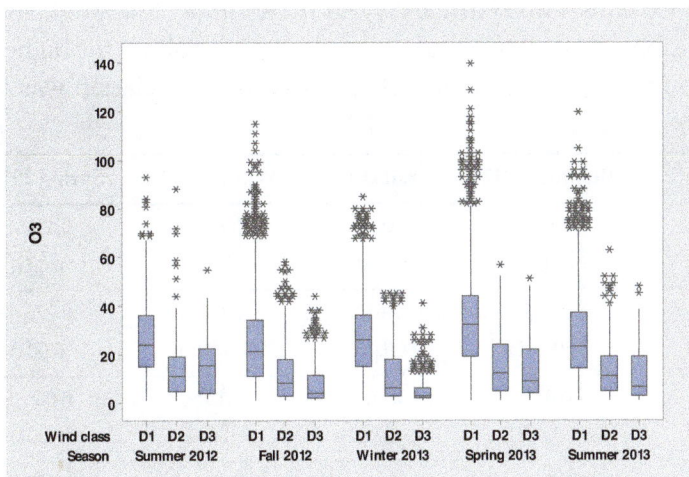

Figure 7. Boxplots of 1-hr averaged O_3 concentrations observed at the Downtown station, classified by prevailing wind direction: D1 (wind blowing from 20 to 120°), D2 (wind blowing from 270 to 360°), and D3 (wind blowing from 180 to 150°)

3.4. Analysis of NO_y levels

Up to this point, our analysis has focused on describing the relationship between NO_x and O_3 production based on the meteorological conditions that foster photochemical reactions. The generalized chemical mechanism that describes this relationship indicates that other reaction products can be of interest in describing the dynamics of O_3 production. In particular, the levels of NO_y provide additional information and serve as an indicator of the degree to which the air masses have been subjected to photochemical processing (aging), thereby providing information on O_3 production.

Table 2 presents the results of the correlation analysis of different chemical and meteorological parameters with NO_y. A correlation coefficient (R^2) of NO_2 with NO_y above 0.78 is noted for months in which O_3 production is not the highest, and between 0.40 and 0.68 for the months with the highest O_3 production. In general, a high percentage of NO_y corresponds to NO_x (Figure 2); thus, the response of one clearly depends on the other.

The difference between NO_y and NO_x (known as NO_z) includes the inorganic and organic nitrates derived from photochemical processing of the NO_x-VOCs mixtures. The typical composition of NO_z is approximately 50%-55% HNO_3, 35%-40% PAN, 1%-5% HNO_2, and a marginal contribution of other nitrates [26, 27]. During the cold months, the production of NO_z can be inhibited by meteorological conditions that do not foster photochemical reactions. Thus, the relative contribution of NO_x to NO_y is the highest, and, therefore, the responses of both species are expected to show high correlation. When conditions favor O_3 production, more NO_z is produced, thereby limiting the contribution of NO_x to the total NO_y. Consequently, the level of correlation drops as the nonlinear photochemistry and sinks of NO_y increase in importance. Of interest is the fact that during spring, the season with the highest peak O_3 1-hr concentrations, the correlation of NO_x species with NO_y is the lowest. A similar response is obtained between CO and NO_y, which can be explained by the typical strong correlation between CO and NO_x emissions, particularly in urban areas that are heavily influenced by mobile-source emissions. With respect to meteorological variables, the highest correlation was with wind speed, and, as in the case of NO_x and CO, the correlation was the lowest during spring.

	Summer 2012	Fall 2012	Winter 2013	Spring 2013	Summer 2013
NO_2 (ppbv)	0.811	0.676	0.819	0.411	0.783
	(< 0.001)	(< 0.001)	(< 0.001)	(< 0.001)	(< 0.001)
NO (ppbv)	0.335	0.610	0.661	0.275	0.615
	(< 0.001)	(< 0.001)	(< 0.001)	(< 0.001)	(< 0.001)
O_3 (ppbv)	0.031	0.106	0.151	0.065	0.070
	(< 0.001)	(< 0.001)	(< 0.001)	(< 0.001)	(< 0.001)
CO (ppmv)	0.764	0.420	0.664	0.227	0.712
	(< 0.001)	(< 0.001)	(< 0.001)	(< 0.001)	(< 0.001)
Solar radiation (kW/m²)	0.011	0.001	0.004	0.001	0.009
	(< 0.001)	(0.639)	(0.052)	(0.165)	(< 0.001)
Temperature (°C)	0.009	0.024	0.022	0.005	0.063
	(0.048)	(< 0.001)	(< 0.001)	(0.002)	(< 0.001)
Wind speed (km/h)	0.252	0.247	0.203	0.112	0.225
	(< 0.001)	(< 0.001)	(< 0.001)	(< 0.001)	(< 0.001)

Table 2. Correlation coefficients (R^2) of parameters monitored by the SIMA downtown station with NO_y (in parenthesis p-values at an α of 0.05)

The results of the correlation analysis suggest a relationship between NO_x and NO_y, but a correlation between NO_y and O_3 is less evident. Figure 2 shows two peaks in the annual cycle of the NO_y monthly average concentration (constructed from daily NO_y average concentrations). One peak occurs during the cold months and is associated with the high levels of NO_x (low NO_z component), while the second peak appears during spring, which is the period with highest photochemical activity (highest O_3 levels); thus, as NO_x is consumed, and its levels decrease and NO_y levels are high, NO_z must be contributing heavily to NO_y. The peaks of NO_y are associated with different conditions, so we looked in greater detail into the seasonal levels of NO_y. Figure 8 depicts the 1-hr NO_y concentrations by season. The highest levels (both of peak and season-average concentrations) occur in winter, followed by fall and spring (when O_3 is the highest), with summer registering the lowest levels. The O_3 levels presented in Figure 3 do not appear to track the NO_y levels.

We then attempted to determine a clearer relationship between O_3 and NO_y by conducting a Principal Components Analysis (PCA) (see Table 3). The first principal component (PC1) has to do with O_3 precursors in the form of primary emissions or intermediate products (NO_x, NO_y, and CO), in line with the results from the descriptive statistics. The second principal component (PC2) covers photochemical O_3 production (O_3 and solar radiation). The third principal component (PC3) refers to a combined meteorological effect that includes horizontal transport (wind direction) and temperature— an effect that was also observed in the descriptive statistics results. The first two principal components represent almost 60% of the variance.

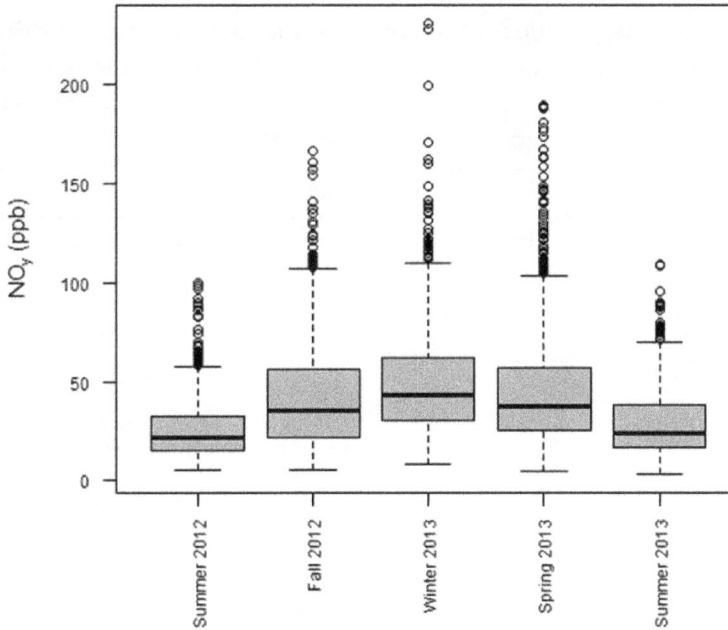

Figure 8. Annual average distribution of NO_y concentrations

Component	PC1	PC2	PC3
NO_y	**0.481**	0.055	−0.009
NO_2	**0.387**	0.087	−0.123
NO	**0.375**	0.029	0.151
O_3	−0.226	**0.447**	0.075
CO	**0.400**	0.110	0.081
Solar radiation	−0.090	**0.497**	0.031
Temperature	−0.197	0.370	**0.411**
Wind velocity	−0.288	0.276	0.084
Wind speed	0.175	0.339	**0.794**
Accumulated variance (%)	0.395	0.595	0.672

Table 3. Principal components analysis results

Given these results, we conducted an additional assessment on the effects of meteorological parameters on O_3 and NO_y levels, which showed the effects of horizontal transport. O_3, temperature, wind speed, and wind direction data were particularly correlated (see Table 4). In addition, complementary polar plots were constructed for NO_y (Figure 9). The highest correlation between O_3 and wind speed occurred during the winter of 2013. In this case, the slope of the linear regression was positive (i.e., as the wind speed increases, O_3 also increases). In a similar way, the highest correlation between O_3 and wind direction occurred during the winter of 2013. This indicates that transport is a major factor responsible for the average O_3 levels observed in the winter in the MMA. The polar plot for NO_y confirms that high levels of NO_y during winter (which are, as discussed, mostly NO_x) are associated with low wind speeds. This relationship between NOy and wind speed would indicate that NO_x is not being processed and translated to high local production of O_3 in the winter.

	Wind speed				
	Summer 2012	Fall 2012	Winter 2013	Spring 2013	Summer 2013
R^2	0.212	0.271	0.350	0.025	0.231
Slope	2.348	3.625	3.342	1.164	2.986
Intercept	6.836	0.937	3.319	35.560	4.850
p-value	< 0.001	< 0.001	< 0.001	< 0.001	< 0.001
	Winddirection				
	Summer 2012	Fall 2012	Winter 2013	Spring 2013	Summer 2013
R^2	0.001	0.022	0.216	0.002	0.010
Slope	−0.005	−0.031	−0.088	0.014	−0.027
Intercept	25.432	24.893	35.313	44.662	28.428
p-value	< 0.001	< 0.001	< 0.001	< 0.001	< 0.001
	Temperature				
	Summer 2012	Fall 2012	Winter 2013	Spring 2013	Summer 2013
R^2	0.454	0.327	0.341	0.344	0.536
Slope	2.834	1.877	2.017	2.031	3.792
Intercept	−57.407	−21.939	−14.324	−5.969	−81.202
p-value	< 0.001	< 0.001	< 0.001	< 0.001	< 0.001

Table 4. O_3 correlations with meteorological parameters (p-values at an α of 0.05)

An additional way to establish the influence of transport on the levels of air pollutants is to analyze the CO-to-NO_y ratio. Values close to 10 are indicative of an influence of local sources, and ratios larger than 100 indicate the influence of regional (remote) sources [11]. Figure 10 depicts the CO-to-NO_y ratios observed at the Downtown station as a function of the season. The average ratio tends to be the inverse with respect to the behavior of the average levels of NO_y, although one main difference exists: the lowest ratios are observed during winter but also during spring, intermediate values are observed during the fall, and the highest average values are registered during the summer. In a few cases, the CO-to-NO_y ratio reached values of 100 or more, and the 95th percentile, in general, was below 56. Thus, air pollution in the MMA during winter appears to be heavily influenced by low wind conditions (Figure 2 indicates the lowest average wind speeds during this season), and poor photochemical activity limits the levels of NO_z and O_3. The O_3 levels that are registered are characteristic of background amounts that are transported to the urban center when wind speeds are sufficiently strong. The spring level is also heavily influenced by local emissions, but meteorological conditions do favor photochemical activity, resulting in production of the highest levels of O_3. Fall would seem to be influenced partly by local emissions and partly by transport. Finally, summer is characterized by the influence of more regional (and aged) emissions. During the first two months of the summer season, the ambient conditions tend to be characterized by high temperatures and deep planetary boundary layers that promote vertical mixing. In addition, conditions favor the mixing of these vertical columns with pollutants that undergo long-range transport.

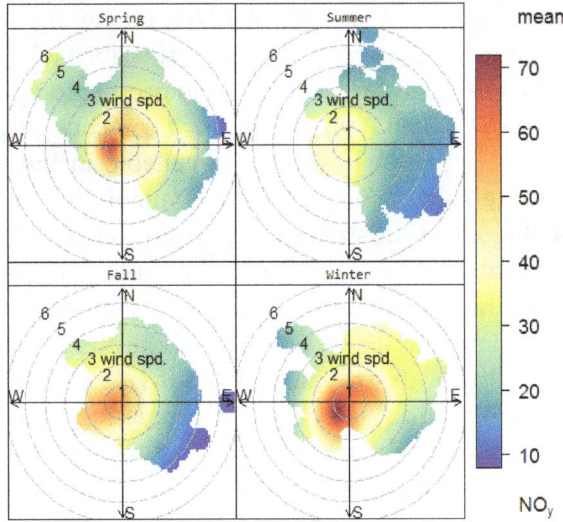

Figure 9. Polar plot for NO_y at the Downtown monitoring station

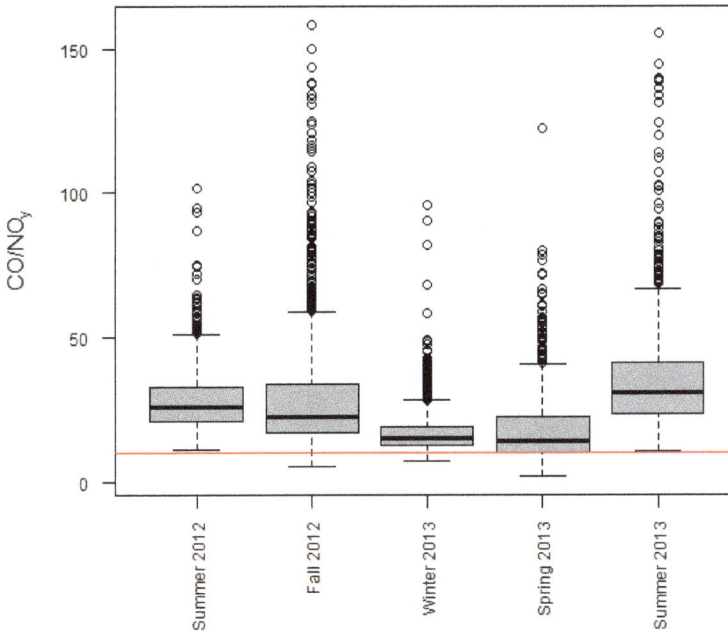

Figure 10. Seasonal variation of CO-to-NO_y ratios (the red line represents a value of 10 for the ratio)

3.5. Weekend-weekday effect

A final aspect of interest in this study was to define if a weekend-weekday effect could be established based on NO_x and NO_y readings. Previous studies have demonstrated that O_3 levels in urban centers can be higher during weekends [28, 29], a condition that might be thought of as counterintuitive, as emissions from anthropogenic sources tend to be lower during weekends. An explanation for the occurrence of this weekend-weekday effect can be obtained by analyzing the photochemical mechanism for the production of O_3. Urban centers tend to be VOC-sensitive; that is, NO_x is in excess and, thus, processes that remove NO_2 from the reactive mixtures are promoted. For example, NO_2 efficiently scavenges $HO^\bullet$ radicals through reaction 9, producing HNO_3, which is a direct sink for NO_x. This limits the reaction of $HO^\bullet$ radicals with VOCs, which, in turn, controls the production pathways for O_3. During weekends, the ratio at which VOCs and NO_x are emitted can change, preferentially reducing the emissions of NO_x. If this happens, less NO_x is available in the atmosphere, reaction 9 becomes less important, more $HO^\bullet$ radicals become available to react with the VOCs, and, thus, O_3 tends to increase. Figure 11 illustrates the distribution of NO_y concentrations by day of the week. At a first glance, NO_y values tend to be quite constant throughout the day. However, Tuesdays and Wednesdays present the lowest peak values, while average levels on Sunday tend to be lower than during the rest of the week. If the average NO_x-to-NO_y ratio is examined, values tend to be lower during the weekends (Figure 12). The NO_x-to-NO_y ratio provides information on the level of photochemical processing of the observed air parcels. Ratios closer to 1 would indicate that most of the NO_y is in the form of NO_x, and thus, it could imply the presence of fresh emissions or the transport of unprocessed emissions. As the ratio drops, the relative contribution of NO_z increases, indicating that an aged (photochemically processed) air parcel is being observed.

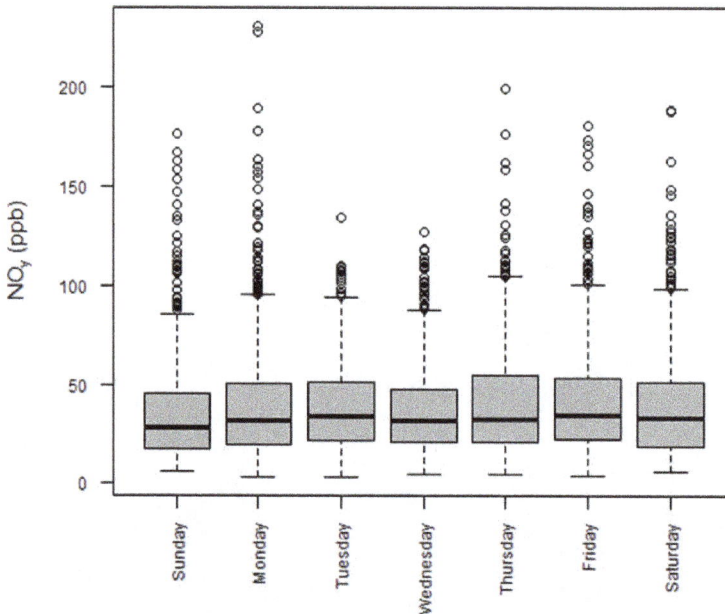

Figure 11. Boxplot for NO_y levels by day of the week

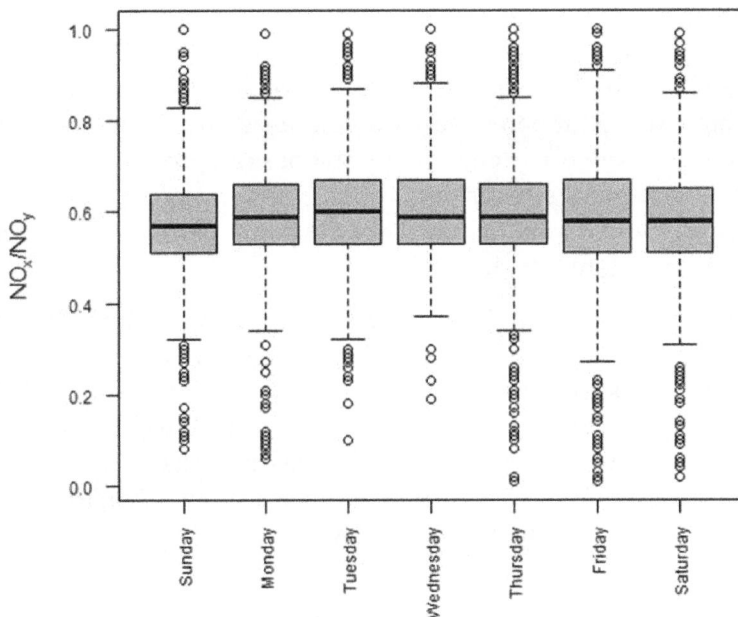

Figure 12. Boxplot for the NO_x-to-NO_y ratio by day of the week

Finally, we conducted an Analysis of Variance (ANOVA) test for the NO_x-to-NO_y ratio by season (see Table 5). For the hot months (spring and summer), lower ratios were observed during weekends compared to weekdays. In fact, the lowest average ratio was obtained during spring, the season with the highest average and peak O_3 concentrations. In contrast, the winter of 2013 showed a marginally higher ratio during the weekends than on the weekdays, but the difference was not statistically significant. This could be the result of an intensive use of fuels during the whole week in response to the low temperatures. For the fall of 2012, this statistical test also indicates no difference between weekdays and weekends. Fall, as the results from previous sections suggest, is influenced by days with high O_3 production and days with very low O_3 levels as a result of the transition from a period with high SR and temperature to the regional rainy season.

Season	Average ± Standard Dev. (Weekdays)	Average ± Standard Dev. (Weekends)	*p-value*	Difference between Weekdays and Weekends
Summer 2012	0.606 ± 0.139	0.582 ± 0.124	0.033	Yes
Fall 2012	0.671 ± 0.141	0.668 ± 0.107	0.657	No
Winter 2013	0.545 ± 0.086	0.549 ± 0.077	0.504	No
Spring 2013	0.553 ± 0.112	0.496 ± 0.142	< 0.001	Yes
Summer 2013	0.586 ± 0.110	0.561 ± 0.112	< 0.001	Yes

Table 5. Results of the ANOVA test for the NO_x-to-NO_y ratio (*p*-values are at an α of 0.05)

4. Conclusions

Observational-based methods have proven valuable for analysis of the chemical interactions that give rise to high air pollution events in urban areas. In particular, the vast amount of information gathered by networks of air quality stations can provide insight into the production of secondary air pollutants such as O_3, as atmospheric conditions change throughout the year. This analysis can be enhanced if complementary observations are also made of species that are not routinely registered, such as NO_y.

In this study, we presented a statistical analysis performed on the air quality and meteorological data registered by the routine air quality stations of the MMA. The analysis included descriptive statistics, regression analysis, correlation analysis, PCA, and ANOVA, along with the interpretation of bivariate polar plots, wind roses, and boxplots. In addition, ratios of NO_x with NO_y and CO with NO_y provided additional information on the level of chemical processing of the air masses traversing the MMA. When used together, these techniques prove to be complementary, thus providing more robust results.

In the MMA, O_3 registers two distinct annual peaks: one in spring and the other in late summer-early fall. The analysis of meteorological conditions and air pollutant levels indicate that the O_3 concentrations during winter would be characteristic of background conditions that get transported to the urban center when wind speeds are sufficiently strong. O_3 production in winter is small because typical conditions that foster photochemical activity are not present, such as relatively high temperatures and strong solar radiation. During this season, NO_y is composed mainly of locally-emitted NO_x, which corroborates the low photochemical activity. That is, the ambient conditions do not favor the catalytic effect of NO_x to produce O_3. Spring is also heavily influenced by local emissions, but meteorological conditions favor photochemical activity, leading to production of the highest levels of O_3. In this season, winds bring background O_3 to the MMA, and precursors traverse the long-axis of the urban core, allowing for chemical processing, injection of fresh emissions, and high photochemical production. Spring registers the highest amount of NO_z, indicating that NO_x is reacting efficiently with VOCs to produce photochemical oxidants, including O_3. Fall would seem to be influenced partly by local emissions and partly by transport. Even though fall has the second highest peak O_3 levels, the average O_3 in this season is relatively low. This can be explained by the fact that the rainy season occurs during this period. Finally, summer would be characterized by the influence of more regional (and aged) emissions. Deep planetary boundary layers are characteristic of this season, allowing mixing and dispersion of air pollutants that result in the lowest NO_y levels of the year. Thus, even though temperature and solar radiation levels could suggest high O_3 production, photochemical activity is limited by transport and mixing effects. Finally, for the hot months (spring and summer), a distinct weekday-weekend effect can be identified, as the NO_x-to-NO_y ratio tends to be higher during weekdays than during weekends. This would indicate changes in NO_x emission rates during the week, which could lead to higher O_3 events during the weekend, in line with the VOC-sensitive condition of the MMA atmosphere that has been suggested by others.

Overall, the analysis of NO_x and NO_y levels with other chemical and meteorological variables as well as the correlation and ratios between NO_x and NO_y provide indicators of the level of photochemical activity that fosters O_3 production.

Acknowledgements

This study was supported by the Mexican Council for Science and Technology (CONACYT) through grant No. CB-2010-1-154122. Additional support was received from Tecnológico de Monterrey through grant no. CAT-186. E. Carrillo appreciates the scholarship received from CONACYT during his research stay (MSc) at Tecnológico de Monterrey.

Author details

Alberto Mendoza* and Edson R. Carrillo

*Address all correspondence to: mendoza.alberto@itesm.mx

School of Engineering and Sciences, Tecnológico de Monterrey, Monterrey, Mexico

References

[1] Seinfeld JH. Urban Air Pollution: State of the Science. Science 1989; 243(4892) 745-752.

[2] Atkinson R. Atmospheric chemistry of VOCs and NOx. Atmospheric Environment 2000; 34(12–14) 2063-2101.

[3] Yang C, Yang H, Guo S, Wang Z, Xu X, Duan X, Kan H. Alternative ozone metrics and daily mortality in Suzhou: The China Air Pollution and Health Effects Study (CAPES). Science of The Total Environment 2012; 426: 83-89.

[4] Ha S, Hu H, Roussos-Ross D, Haidong K, Roth J, Xu X. The effects of air pollution on adverse birth outcomes. Environmental Research 2014; 134: 198-204.

[5] Vlachokostas C, Nastis SA, Achillas C, Kalogeropoulos K, Karmiris I, Moussiopoulos N, Chourdakis E, Banias G, Limperi N. Economic damages of ozone air pollution to crops using combined air quality and GIS modelling. Atmospheric Environment 2010; 44(28) 3352-3361.

[6] Feng Z, Sun J, Wan W, Hu E, Calatayud V. Evidence of widespread ozone-induced visible injury on plants in Beijing, China. Environmental Pollution 2014; 193: 296-301.

[7] Fang Y, Naik V, Horowitz LW, Mauzerall DL. Air pollution and associated human mortality: the role of air pollutant emissions, climate change and methane concentration increases from the preindustrial period to present. Atmospheric Chemistry and Physics 2013; 13(3) 1377-1394.

[8] Seinfeld JH, Pandis SN. Atmospheric Chemistry and Physics: From Air Pollution to Climate Change. 2nd ed. Hoboken, New Jersey: John Wiley & Sons.; 2006.

[9] Warnek P. Chemistry of the Natural Atmosphere. International Geophysics Series. New York: Academic Press. 757; 1988.

[10] Russell A, Dennis R. NARSTO critical review of photochemical models and modeling. Atmospheric Environment 2000; 34(12–14) 2283-2324.

[11] Kleanthous S, Vrekoussis M, Mihalopoulos N, Kalabokas P, Lelieveld J. On the temporal and spatial variation of ozone in Cyprus. Science of The Total Environment 2014; 476–477: 677-687.

[12] Nishanth T, Praseed KM, Kumar MKS, Valsaraj KT. Influence of ozone precursors and PM_{10} on the variation of surface O_3 over Kannur, India. Atmospheric Research 2014; 138: 112-124.

[13] Iqbal MA, Kim K-H, Shon Z-H, Sohn J-R, Jeon E-C, Kim Y-S, Oh J-M. Comparison of ozone pollution levels at various sites in Seoul, a megacity in Northeast Asia. Atmospheric Research 2014; 138: 330-345.

[14] Zhang Y, Mao H, Ding A, Zhou D, Fu C. Impact of synoptic weather patterns on spatio-temporal variation in surface O_3 levels in Hong Kong during 1999–2011. Atmospheric Environment 2013; 73: 41-50.

[15] Alghamdi MA, Khoder M, Harrison RM, Hyvärinen AP, Hussein T, Al-Jeelani H, Abdelmaksoud AS, Goknil MH, Shabbaj II, Almehmadi FM, Lihavainen H, Kulmala M, Hämeri K. Temporal variations of O_3 and NO_x in the urban background atmosphere of the coastal city Jeddah, Saudi Arabia. Atmospheric Environment 2014; 94: 205-214.

[16] Jenkin ME. Investigation of the impact of short-timescale NO_x variability on annual mean oxidant partitioning at UK sites. Atmospheric Environment 2014; 90: 43-50.

[17] Hassan IA, Basahi JM, Ismail IM, Habeebullah TM. Spatial Distribution and Temporal Variation in Ambient Ozone and Its Associated NO_x in the Atmosphere of Jeddah City, Saudi Arabia. Aerosol and Air Quality Research 2013; 13 1712-1722.

[18] González-Santiago O, Badillo-Castañeda CT, Kahl JDW, Ramírez-Lara E, Balderas-Renteria I. Temporal Analysis of PM_{10} in Metropolitan Monterrey, México. Journal of the Air & Waste Management Association 2011; 61(5) 573-579.

[19] Sistema Integral de Monitoreo Ambiental. Estadística SIMA: Gobierno del Estado de Nuevo León. http://www.nl.gob.mx/?P=med_amb_mej_amb_sima_estadisti (accessed 1 Oct 2013).

[20] Sierra A, Vanoye AY, Mendoza A. Ozone sensitivity to its precursor emissions in northeastern Mexico for a summer air pollution episode. Journal of the Air & Waste Management Association 2013; 63(10) 1221-1233.

[21] Green J, Sánchez S. Air Quality In Latin America: An Overview. Washington, DC: Clean Air Institute; 2013. http://www.cleanairinstitute.org/calidaddelaireamericalatina/ cai-report-english.pdf (accessed 22 September 2014).

[22] Mancilla Y. $PM_{2.5}$ Source Apportionment using Partial Least Squares Based on Organic Molecular Markers: Monterrey Metropolitan Area, Mexico. PhD thesis. Tecnológico de Monterrey; 2013.

[23] SEMARNAT, Gobierno del Estado de Nuevo León. Programa de Gestión para Mejorar la Calidad del Aire del Área Metropolitana de Monterrey 2008-2012. 2008. http://www.semarnat.gob.mx/archivosanteriores/temas/gestionambiental/calidaddelaire/ Documents/Calidad%20del%20aire/Proaires/ProAires_Vigentes/6_ProAire%20AMM %202008-2012.pdf (accessed 23 September 2014).

[24] Rodwell MJ, Hoskins BJ. Subtropical Anticyclones and Summer Monsoons. Journal of Climate 2001; 14(15) 3192-3211.

[25] Sillman S. The relation between ozone, NO_x and hydrocarbons in urban and polluted rural environments. Atmospheric Environment 1999; 33(12) 1821-1845.

[26] Luke WT, Kelley P, Lefer BL, Flynn J, Rappenglück B, Leuchner M, Dibb JE, Ziemba LD, Anderson CH, Buhr M. Measurements of primary trace gases and NO_y composition in Houston, Texas. Atmospheric Environment 2010; 44(33) 4068-4080.

[27] Luecken DJ, Tonnesen GS, Sickles IJE. Differences in NO_y speciation predicted by three photochemical mechanisms. Atmospheric Environment 1999; 33(7) 1073-1084.

[28] Wang YH, Hu B, Ji DS, Liu ZR, Tang GQ, Xin JY, Zhang HX, Song T, Wang LL, Gao WK, Wang XK, Wang YS. Ozone weekend effects in the Beijing–Tianjin–Hebei metropolitan area, China. Atmospheric Chemistry and Physics 2014; 14 2419-2429.

[29] Warneke C, de Gouw JA, Edwards PM, Holloway JS, Gilman JB, Kuster WC, Graus M, Atlas E, Blake D, Gentner DR, Goldstein AH, Harley RA, Alvarez S, Rappenglueck B, Trainer M, Parrish DD. Photochemical aging of volatile organic compounds in the Los Angeles basin: Weekday-weekend effect. Journal of Geophysical Research: Atmospheres 2013; 118(10) 5018-5028.

3

PM_{10} Time Series Analysis Through Geostatistical Techniques

Claudia Cappello, Sabrina Maggio,

Daniela Pellegrino and Donato Posa

Additional information is available at the end of the chapter

1. Introduction

Particulate matter (PM) is an air pollutant comes from vehicular traffic, industrial activities and street dust, or from the atmosphere, by transformation of the gaseous emissions. In recent years the interest in the health effects of this pollutant have increased, since high concentration levels in urban area have been measured.

Several studies suggest an association between fine particulate air pollution and the increase of the mortality rate [1]. In particular, PM up to 10 micrometers in size (PM_{10}) could cause negative health effects such as respiratory illness or cardiovascular problems. Hence, the analysis of temporal evolution of this pollutant could be useful in decision-making process for environmental policy.

Typically, in time series analysis, the Box-Jenkins methodology is widely applied and the autocorrelation function (ACF) is used as a standard exploratory tool to identify the model structure [3, 4]. In this context, the use of geostatistical techniques could also be convenient, nevertheless these techniques are usually applied to analyze, through the variogram, spatial relationships among sample data measured at some locations in a domain and to predict the corresponding spatial phenomena [6, 18, 22, 29]. In particular, the variogram could represent a complementary exploratory tool for assessing stationarity in time series [2, 19] and it has the considerable advantage that it is defined in much wider circumstances than the autocovariance and the autocorrelation. Moreover, this analytical tool is appropriate to identify trends and periodicity exhibited by the data and to obtain kriging predictions of the variable under study, either for temporal intervals with missing values (interpolation mode) and in time points after the last available data (extrapolation mode).

Different studies have suggested the use of geostatistical methods in time domain [7, 19]. In particular, De Iaco et al. [12] illustrated the role of variogram in this context for different purposes.

The aim of this paper is to analyze PM_{10} air pollution in an area of South Italy characterized by high levels of industrial emissions and vehicular traffic, through geostatistical techniques.

Thus, after a brief review on stochastic processes and geostatistical methods in time series analysis, the temporal evolution of PM_{10} daily concentrations, for the period 2010-2013 has been assessed. After the identification of trend and periodicity, the reconstruction of the analyzed time series by estimation of missing values has been discussed, and predictions of PM_{10} daily concentrations at some unsampled points have been produced. Moreover, the probability distributions of the variable under study have been estimated for future time points.

For interpolation and prediction purposes, a modified version of *GSLib* kriging routine has been used.

2. Theoretical framework

In time series analysis the observed values of a variable for different time points or intervals can be reasonably considered as a finite realization of a real-valued random process, denoted with $\{X_t, t \in T \subseteq \mathbb{R}\}$.

Besides the common second-order moments used to describe the random process $\{X_t, t \in T\}$, such as the autocovariance function and the autocorrelation function, the variogram can also be considered and even preferred with respect to covariance function [10, 19].

Given a stochastic process $\{X_t, t \in T\}$ over a temporal domain $T \subseteq \mathbb{R}$, the corresponding variogram is defined as follows

$$\gamma(t, t + h_t) = 0.5 Var\left[X_t - X_{t+h_t}\right], \qquad t, t + h_t \in T. \tag{1}$$

Note that a function $\gamma(\cdot)$ is a variogram if and only if it is conditionally strictly negative definite [23].

As known in the literature [3–5], time series analysis is based on the theory of stationary processes. It is worth highlighting that the second-order stationarity implies the intrinsic stationarity, but the converse is not true [18, 22].

In particular, the stochastic process $\{X_t, t \in T\}$ is intrinsically stationary if its variogram $\gamma(t, t + h_t)$ depends solely on the temporal lag h_t and the expected value of the difference $(X_t - X_{t+h_t})$ is constant.

The variogram, widely used in geostatistical context, could be applied efficiently in time series analysis [14, 15], since

- it can describe a wider class of stochastic processes, i.e. the class of intrinsic stochastic processes, which includes the class of second-order stationary stochastic processes,

- its estimation does not require the knowledge of the expected value of the associated stochastic process,
- it is appropriate to identify trend and periodicity exhibited by data,
- it can be used for prediction purposes.

Regarding this last aspect, geostatistical techniques provide different parametric and nonparametric prediction methods, among these the sample and ordinary kriging, the universal kriging and the indicator kriging. Further details can be found in the specialized literature [7, 12, 19]. Thus, the estimation of the unknown value x_t of the stochastic process $\{X_t, t \in T\}$, using the data observed in the past (extrapolation mode), or the data observed before and after the time point t (interpolation mode) can be easily supported by geostatistical tools.

In the following, the ordinary kriging method and the indicator kriging approach are briefly reviewed, since these geostatistical tools are used for analyzing the variable under study.

Let $\widehat{X}_t$ the linear predictor of the intrinsic stationary process $\{X_t, t \in T\}$:

$$\widehat{X}_t = \sum_{i=1}^{n} \lambda_i(t) X_{t_i}, \tag{2}$$

where $\lambda_i(t)$, $i = 1, 2, \ldots, n$, are unknown real coefficients and X_{t_i} are random variables of the process X at the sampled time points t_i. The unknown weights $\lambda_i(t)$, $i = 1, 2, \ldots, n$, of (2) are obtained by solving the following kriging system

$$\begin{bmatrix} \gamma_{11} & \cdots & \gamma_{1n} & -1 \\ \gamma_{21} & \cdots & \gamma_{2n} & -1 \\ \vdots & \ddots & \vdots & \vdots \\ \gamma_{n1} & \cdots & \gamma_{nn} & -1 \\ 1 & \cdots & 1 & 0 \end{bmatrix} \begin{bmatrix} \lambda_1 \\ \lambda_2 \\ \vdots \\ \lambda_n \\ \mu \end{bmatrix} = \begin{bmatrix} \gamma_{10} \\ \gamma_{20} \\ \vdots \\ \gamma_{n0} \\ 1 \end{bmatrix}, \tag{3}$$

where $\gamma_{ij} = 0.5\,Var(X_{t_i} - X_{t_j})$, $\gamma_{i0} = 0.5Var(X_{t_i} - X_t)$, μ is the Lagrange multiplier. If γ is conditionally strictly negative definite, then the above system presents one and only one solution.

The ordinary kriging [22] requires only the knowledge of the variogram model and it is used when the expected value of the process is constant and unknown. Since the kriging system can be expressed in terms of the variogram, as in (3), the kriging predictor can be used even when the stochastic process under study satisfies the intrinsic hypothesis. Moreover, using a predictor based on a variogram, rather than on a covariance, avoids the estimation of the expected value, if this last is unknown.

The usefulness of geostatistical techniques in time series analysis can be appreciated through nonparametric estimation of the variable under study.

The kriging approach, based on the knowledge of variogram, leads naturally to nonparametric estimation [17]. Indicator kriging is a nonparametric approach to estimate the posterior cumulative distribution function (c.d.f.) of the variable under study at an unsampled point [16, 25, 26].

In this context, given the observed time series x_{t_i}, t_i, $i = 1, 2, ..., n$, the conditional probability $Prob\{X_t \leq x | \mathcal{H}_n\}$, with $\mathcal{H}_n = \{x_{t_i}, t_i, i = 1, 2, ..., n\}$, is interpreted as conditional expectation of an indicator random field $I(t; x)$ [27], that is

$$Prob\{X_t \leq x | \mathcal{H}_n\} = E\left[I(t; x | \mathcal{H}_n)\right]$$

where

$$I(t; x) = \begin{cases} 1, & if \ X_t \leq x \\ 0, & if \ X_t > x. \end{cases}$$

In the case study presented hereafter, ordinary kriging and indicator kriging are applied for interpolation and prediction purposes of an environmental variable. Note that a *GSLib* routine for kriging, named "KT3DP" [12], has been used in order to define appropriate temporal search neighborhoods in presence of periodicity, since environmental time series, such as the ones for air pollution data, usually are characterized by a periodic behavior. Hence, the use of periodic and nonperiodic variogram models have been proposed through two different approaches:

- the periodic component has been factored out using the moving average method [5] and nonperiodic variogram model has been fitted;

- the periodicity has been retained and described by a periodic variogram model.

3. PM_{10} time series

In the present case study, the analysis of daily concentrations of PM_{10} ($\mu g/m^3$), measured at one of the monitoring stations of Brindisi district during the period 2010-2013, has been conducted through geostatistical techniques.

These data have been collected by the Environmental Protection Agency of Apulian region (ARPA Puglia) which controls the air quality of urban, suburban, and industrialized areas of the region.

Note that PM_{10} monitoring stations are classified in the following three categories:

- traffic stations, located in areas with heavy traffic;

- industrial stations, located close to industrialized areas;

- background stations, located in peripheral areas.

The analyzed station, named "Torchiarolo" is located in the municipality of Torchiarolo (Brindisi district), as shown in Fig. 1. It is classified as industrial station, since it is strictly close to an industrial site, i.e. the thermoelectric power station "Enel-Federico II" in Cerano (Brindisi district).

Figure 1. "Torchiarolo" monitoring station belonging to the Environmental Protection Agency of Apulian region (ARPA Puglia)

3.1. Exploratory Data Analysis

In order to assess the statistical properties of PM_{10} measured at the "Torchiarolo" station in the period 2010-2013, an exploratory data analysis has been performed. Some results are shown in Tab. 1.

Year	Min	Max	Mean	Standard Deviation	Number of exceedances
2010	8	114	35.10	20.09	67
2011	8	147	36.08	21.57	66
2012	10	108	32.83	17.01	49
2013	8	146	35.83	22.71	61
2010-2013	8	147	34.97	20.47	243

Table 1. Descriptive statistics of PM_{10} ($\mu g/m^3$), measured in the period 2010-2013 at the "Torchiarolo" monitoring station

According to the National Law concerning the human health protection, PM_{10} daily average concentrations cannot be greater than 50 $\mu g/m^3$ for more than 35 times per year. During the period under study, the PM_{10} daily values exceeded the threshold 243 times; in addition, the station has measured more than 35 exceedances per year.

The box plot in Fig. 2 shows that the observed time series is characterized by a seasonal component. During summertime, particle pollution shows lower levels compared to those recorded during wintertime; in particular, in summertime, PM_{10} doesn't exceed the limit value fixed by the National Law.

On the other hand, in wintertime changes in the lower layer of the troposphere determine PM_{10} stagnation. Hence, high levels of this pollutant are recorded.

In the following sections, the study of the temporal evolution of PM_{10} at the analyzed station has been conducted by performing

- structural analysis,
- estimation of some consecutive values assumed as missing,
- prediction of PM_{10} daily averages,
- estimation of the c.d.f. of PM_{10} daily concentrations at some unsampled time points.

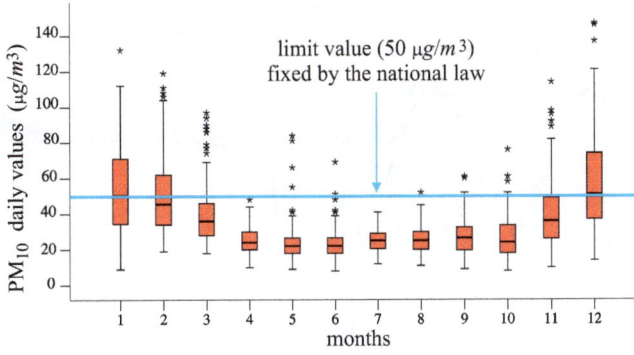

Figure 2. Box plot of PM_{10} daily concentrations, grouped by month, and limit value fixed by the National Decree (Decree-law 60/2002)

4. Structural Analysis

As previously pointed out, the variogram can describe a wider class of stochastic processes, that is the class of intrinsic stochastic processes and is usually preferred to the use of the covariance function.

In structural analysis, before modeling the temporal correlation described by the variogram, its estimation from data is required. The following classical estimator [9] is often used:

$$\hat{\gamma}(r_t) = \frac{1}{2\,|M(r_t)|} \sum_{M(r_t)} \left[X_{t+h_t} - X_t\right]^2 \tag{4}$$

where r_t is the temporal lag, $M(r_t) = \{t + h_t \in H$ and $t \in H$, such that $\|r_t - h_t\| < \delta_t\}$, δ_t is the tolerance, H is the set of data at different time points (not necessarily equally-spaced) and $|M(r_t)|$ is the cardinality of this set.

In the present case study the variogram has been used as an exploratory tool to assess stationarity and periodicity. In particular, sample temporal variogram for PM_{10} daily observations, shown in Figure 3-a), reproduces the seasonal behavior of the variable under study, which presents an annual periodicity at 365 days. In equation (5) the analytic expression of the periodic variogram model, fitted to the sample variogram for the observed values, is proposed:

$$\gamma(h_t) = 265\,Exp(|h_t|; 10) + 130\,Cos(|h_t|; 365); \tag{5}$$

where $Exp(\cdot)$ and $Cos(\cdot)$ are the exponential and the cosine variogram models [29], respectively.

On the other hand, since the variable under study, is characterized by periodicity, this seasonal component could be factored out. Moving average and monthly averages techniques have been applied in order to obtain PM_{10} residuals. Note that the FORTRAN program "REMOVE" [11] has been used to apply moving average techniques.

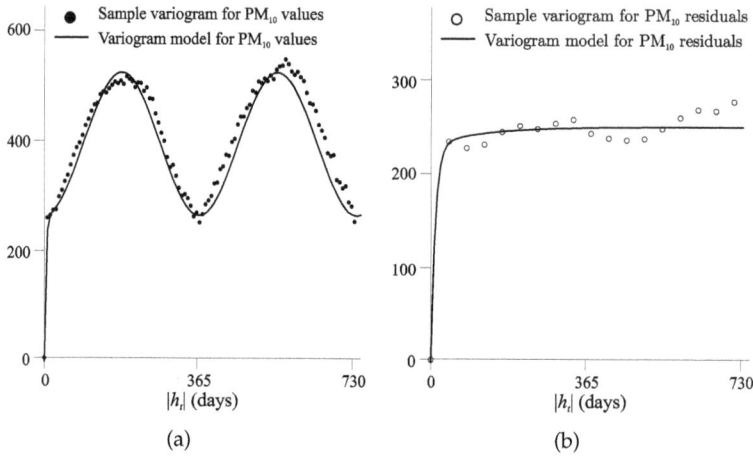

Figure 3. Sample temporal variograms and fitted models. (a) Variogram for PM_{10} daily concentrations (b) Variogram for PM_{10} residuals

The sample variogram of the residuals has been computed and modelled and the following nonperiodic variogram model has been chosen:

$$\tilde{\gamma}(h_t) = 348\, Exp(|h_t|; 30) + 30\, Exp(|h_t|; 365); \tag{6}$$

where $Exp(\cdot)$ is the exponential variogram model [8].

The sample temporal variogram for PM_{10} daily residuals and the corresponding nonperiodic fitted model (6) are illustrated in Fig. 3-b).

In both cases (original data and residuals), the behavior of the variogram functions near the origin is assumed to be linear with no nugget effect.

The goodness of variogram models (5) and (6) has been evaluated through cross-validation, which allows the estimation for PM_{10} daily concentrations and PM_{10} residuals, respectively, at all data points. Figure 4 shows the scatter plots of PM_{10} observed values (a) and PM_{10} residuals (b) towards the corresponding estimated values. The high values of the linear correlation coefficients (0.783 and 0.780, respectively) confirm the goodness of the above fitted models.

It is important to point out that the variogram model (5) has been validated using a modified version of the *GSLib* program "KT3D" [13], named "KT3DP". This program has been developed in order to properly define the neighborhood, i.e. the subset of available data used in the kriging system.

By taking into account the main features of the analyzed pollutant and its temporal behavior (periodicity at 365 days), the kriging routine has been modified and the value at time t is estimated by considering data observed

- at the two adjacent time points $(t-1)$ and $(t+1)$,
- at the same day of the year before and/or later, $(t-d)$ and $(t+d)$, with $d = 365$ and some days before and/or later, $(t-d\pm k)$ and $(t+d\pm k)$, with $k = 1, 2, 3$,

- at the same day of two years before and/or later, $(t - 2d)$ and $(t + 2d)$, with $d = 365$ and some days before and/or later, $(t - 2d \pm k)$ and $(t + 2d \pm k)$, with $k = 1, 2, 3,$

up to a maximum number of eight values.

The variogram model (6), which describes the temporal correlation for PM_{10} residuals, has been validated using the *GSLib* program "KT3D".

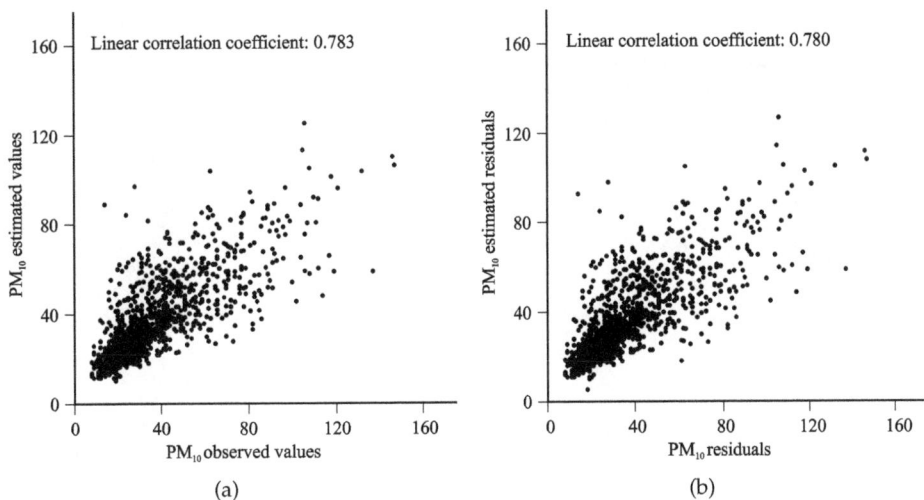

Figure 4. Scatter plots between observed and estimated values. (a) Diagram of PM_{10} daily concentrations towards the estimated ones (b) Diagram of PM_{10} residuals towards the estimated ones

5. Estimation of missing values

In this section the reconstruction of PM_{10}, by using the kriging technique, has been discussed [20, 21, 30, 31].

The reconstruction of temporal data is required if a time series is incomplete. This problem could be due to a malfunction of the monitoring station or the presence of invalid data.

With this aim, six consecutive PM_{10} values from the 12th to the 17th of June 2011, have been considered as missing, both for the observed time series with a 365-day periodic behavior and the deseasonalized values.

Kriging daily estimations for these missing values have been obtained using, alternatively

1. the periodic variogram model (5), which describes the temporal correlation for PM_{10} daily concentrations,

2. the nonperiodic variogram model (6), which describes the temporal correlation for PM_{10} daily residuals.

Since the time series of the observed values is characterized by a periodic behavior, *GSLib* routine "KT3DP", properly modified in order to define an appropriate neighborhood, has been used with the aim to estimate PM_{10} daily measurements.

On the other hand, for the deseasonalized time series, PM_{10} residuals have been estimated by the original version of "KT3D". Finally the periodic component, previously estimated by the moving average and monthly averages techniques, has been added to the estimated residuals, in order to obtain estimates of PM_{10} daily concentrations.

Time series of estimated missing values, obtained with the periodic variogram model (5) and the nonperiodic variogram model (6), are shown in Fig. 5, together with the time series of PM_{10} values, observed on June 2011.

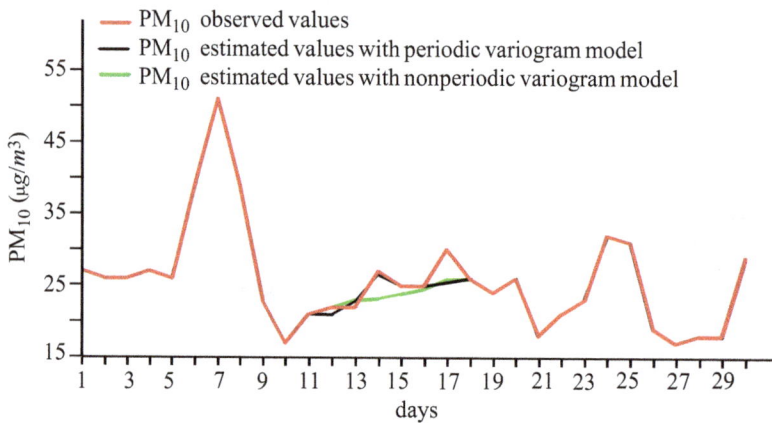

Figure 5. Time plot of PM_{10} estimated missing values and PM_{10} daily concentrations ($\mu g / m^3$), from the 12th to the 17th of June 2011

In order to test the validity of the estimation procedure, the linear correlation coefficients have been computed. In particular, the linear correlation coefficient between the PM_{10} observed values and the corresponding estimates, obtained with the periodic variogram model, is equal to 0.805. On the other hand, the linear correlation coefficient between the residuals and the corresponding estimates, obtained with the nonperiodic variogram model, is equal to 0.831. These results confirm the goodness of the kriging technique as estimator of missing values.

In Table 2 some results of estimation procedure are shown. Note that the mean value of the kriging standard error is lower if the periodic variogram model is used, compared with the kriging results based on the nonperiodic variogram model.

Therefore, the flexibility of kriging to reconstruct the time series has been demonstrated even when the periodic component is not factored out and the temporal correlation is described by a periodic variogram model.

6. Prediction of PM_{10} values

In this section, predictions for the variable under study in time points after the last available data are discussed [23, 24, 28].

The periodic variogram model (5) of PM_{10} concentrations and the nonperiodic variogram model (6) of PM_{10} residuals, have been used in order to predict six time points after the last available data, i.e. the 31st of December 2013. In particular, kriging predictions have

June 2011	PM_{10} value	PM_{10} Est. value[a]	Est. Error[a]	Est. value[b]	Est. Error[b]
12th	22	20.970	-1.030	22.029	0.029
13th	22	22.821	0.821	22.940	0.940
14th	27	26.437	-0.563	23.178	-3.822
15th	25	24.867	-0.133	23.878	-1.122
16th	25	24.926	-0.074	24.473	-0.527
17th	30	25.466	-4.534	25.825	-4.175
Mean values	21.167	24.248	-0.919	23.720	-1.446

[a] Results obtained by using the periodic variogram model (5) [b] Results obtained by using the nonperiodic variogram model (6)

Table 2. Kriging estimations of a sequence of 6 missing values, from the 12th to the 17th of June 2011 and corresponding errors for periodic and nonperiodic variogram models

been computed for the period ranging from the 1st to the 6th of January 2014, by using, alternatively

1. the available data, the variogram model (5) and the modified *GSLib* routine "KT3DP" which builds the searching neighborhood taking into account the periodicity exhibited by the data,

2. the deseasonalized PM_{10} observations, the variogram model (6) and the original *GSLib* routine "KT3D" which produces PM_{10} predicted residuals at which the diurnal component of the day before has been added to obtain predictions of PM_{10} daily concentrations.

In Fig. 6, the time series of PM_{10} daily concentrations measured from the 9th of December

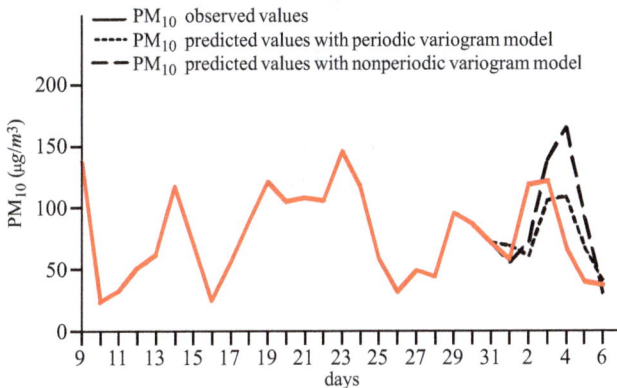

Figure 6. Time plot of PM_{10} predicted values and PM_{10} daily concentrations ($\mu g/m^3$), from the 1st to the 6th of January 2014

2013 to the 6th of January 2014 is shown together with the predicted PM_{10} values for the period ranging from the 1st to the 6th of January 2014. Note that the kriging procedure using the nonperiodic variogram model (6) related to PM_{10} residuals has produced overestimates of the pollution levels.

Moreover, in Table 3 some results of the performance of the prediction procedure are presented. The mean value of the kriging standard error is lower if the periodic variogram model is used, compared with the case of kriging based on the nonperiodic variogram model.

January 2014	PM_{10} Obs. value	PM_{10} Est. value[a]	Est. Error[a]	Est. value[b]	Est. Error[b]
1st	58	69.415	11.415	55.774	-2.226
2nd	119	61.031	-57.969	72.985	-46.015
3rd	122	106.051	-15.949	137.731	15.731
4th	67	108.397	41.397	164.909	97.909
5th	40	67.545	27.545	91.090	51.090
6th	37	41.498	4.498	30.738	-6.262
Mean values	73.833	75.656	1.823	92.205	18.371

[a] Results obtained by using the periodic variogram model (5)   [b] Results obtained by using the nonperiodic variogram model (6)

Table 3. Kriging predictions of a sequence of six days, from the 1st to the 6th of January 2014 and corresponding errors for periodic and nonperiodic variogram models

It is important to highlight that in the period 1-5 January 2014 predicted values greater than 50 $\mu g/m^3$ (i.e. the limit value fixed by the National Law) have been obtained.

Note that in the period 1-4 January 2014, PM_{10} values greater than this threshold have been measured. On the other hand, at day 5th, the kriging procedure produces overestimate of the variable under study.

7. Estimation of the c.d.f.

For a given time series of PM_{10}, it might be useful to estimate the probability that the variable under study exceeds a fixed limit, so that appropriate and prompt solutions might be adopted if necessary.

In this section, estimation of c.d.f. of PM_{10} daily concentrations ($\mu g/m^3$) has been conducted.

In particular, the c.d.f. of PM_{10} at unsampled time points has been estimated by indicator kriging [17].

Six threshold values for PM_{10} (22, 35, 50, 78, 98, and 108 $\mu g/m^3$) have been properly chosen, and six indicator variables according to the fixed thresholds have been defined as follows

- $I_1(t;22) = \begin{cases} 1, & if \ PM_{10} \leq 22 \\ 0, & if \ otherwise \end{cases}$

- $I_2(t;35) = \begin{cases} 1, & if \ PM_{10} \leq 35 \\ 0, & if \ otherwise \end{cases}$

- $I_3(t;50) = \begin{cases} 1, & if \ PM_{10} \leq 50 \\ 0, & if \ otherwise \end{cases}$

- $I_4(t;78) = \begin{cases} 1, & if \ PM_{10} \leq 78 \\ 0, & if \ otherwise \end{cases}$

- $I_5(t;98) = \begin{cases} 1, & if \ PM_{10} \leq 98 \\ 0, & if \ otherwise \end{cases}$

- $I_6(t;108) = \begin{cases} 1, & if \ PM_{10} \leq 108 \\ 0, & if \ otherwise \end{cases}$

with $t \in T$. Note that indicator data are equal to 1 if the values of the variable under study are not greater than the considered threshold and they are equal to 0 otherwise. For each threshold, the temporal indicator variogram has been computed and modelled (Figs. 7, 8).

(a) $x_1 = 22\,\mu g/m^3$

(b) $x_1 = 22\,\mu g/m^3$

(c) $x_2 = 35\,\mu g/m^3$

(d) $x_2 = 35\,\mu g/m^3$

(e) $x_3 = 50\,\mu g/m^3$

(f) $x_3 = 50\,\mu g/m^3$

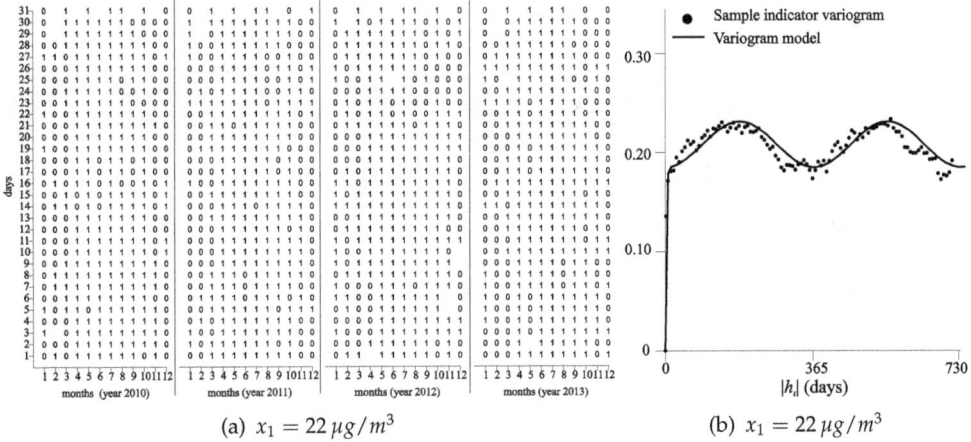

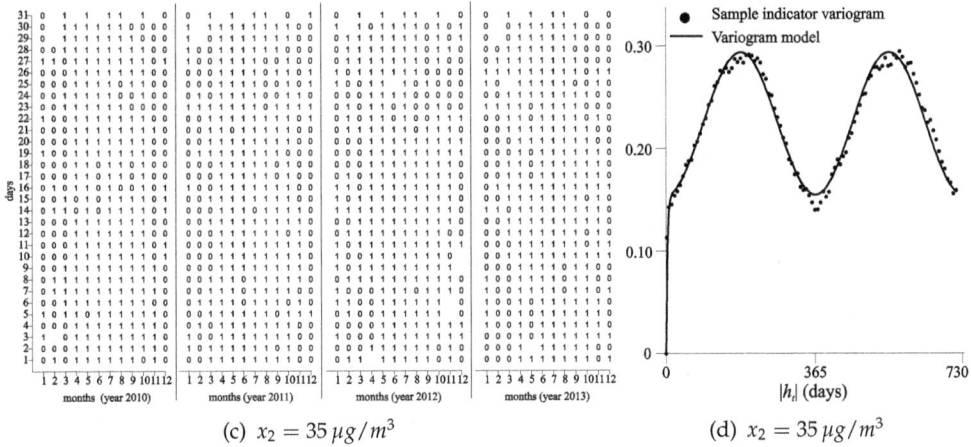

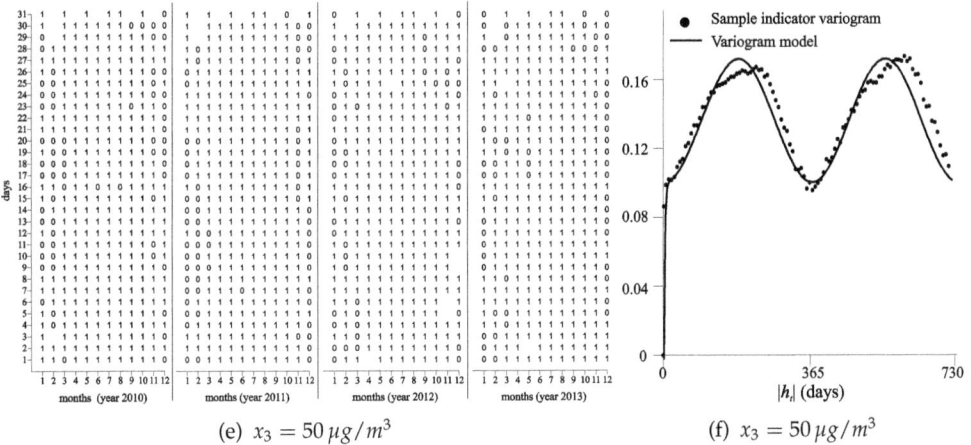

Figure 7. Indicator maps of PM_{10} daily concentrations and their sample indicator variograms with the fitted models, for three threshold values. (a) Indicator map and (b) sample variogram indicator for the threshold $x_1 = 22\,\mu g/m^3$. (c) Indicator map and (d) sample variogram indicator for the threshold $x_2 = 35\,\mu g/m^3$. (e) Indicator map and (f) sample variogram indicator for the threshold $x_3 = 50\,\mu g/m^3$

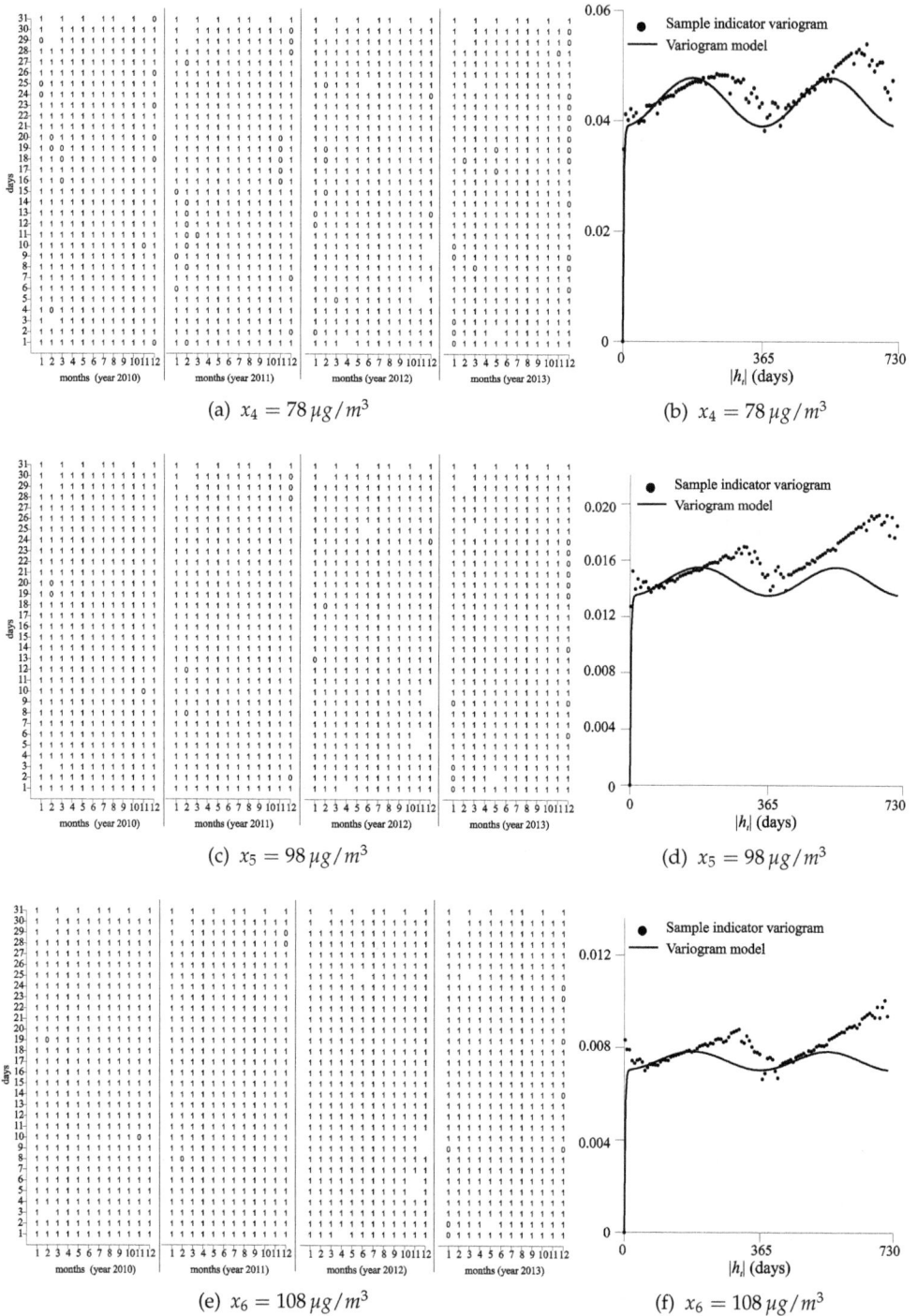

(a) $x_4 = 78\,\mu g/m^3$

(b) $x_4 = 78\,\mu g/m^3$

(c) $x_5 = 98\,\mu g/m^3$

(d) $x_5 = 98\,\mu g/m^3$

(e) $x_6 = 108\,\mu g/m^3$

(f) $x_6 = 108\,\mu g/m^3$

Figure 8. Indicator maps of PM_{10} daily concentrations and their sample indicator variograms with the fitted models, for three threshold values. (a) Indicator map and b) variogram for the threshold $x_4 = 78\,\mu g/m^3$. (c) Indicator map and (d) variogram for the threshold $x_5 = 98\,\mu g/m^3$. (e) Indicator map and (f) variogram for the threshold $x_6 = 108\,\mu g/m^3$

In particular the following models have been fitted

- $\gamma_{I_1}(h_t; 22) = 0.185 \, Exp(|h_t|; 10) + 0.023 \, Cos(|h_t|; 365),$

- $\gamma_{I_2}(h_t; 35) = 0.15 \, Exp(|h_t|; 10) + 0.07 \, Cos(|h_t|; 365),$

- $\gamma_{I_3}(h_t; 50) = 0.102 \, Exp(|h_t|; 10) + 0.036 \, Cos(|h_t|; 365),$

- $\gamma_{I_4}(h_t; 78) = 0.039 \, Exp(|h_t|; 10) + 0.0004 \, Cos(|h_t|; 365),$

- $\gamma_{I_5}(h_t; 98) = 0.013 \, Exp(|h_t|; 10) + 0.001 \, Cos(|h_t|; 365),$

- $\gamma_{I_6}(h_t; 108) = 0.007 \, Exp(|h_t|; 10) + 0.0004 \, Cos(|h_t|; 365).$

Thus, the c.d.f.s corresponding to six different unsampled time points, i.e. the days 1-6 of January 2014, have been estimated by using the "KT3DP" routine.

For each day of interest, the c.d.f. has been estimated by solving as many kriging systems as the number of threshold values considered. For each threshold, the corresponding indicator variogram model has been used for the kriging procedure.

Figure 9 shows the c.d.f.s estimated at days 1-6 of January 2014. It is clear that the probability of not exceeding a fixed threshold increases gradually from the 1st to the 6th of January 2014. For example, the estimated probability that PM_{10} concentrations, on the 1st of January 2014, do not exceed 22 $\mu g/m^3$ is lower than the estimated probability on the 3rd or the 6th of the considered month.

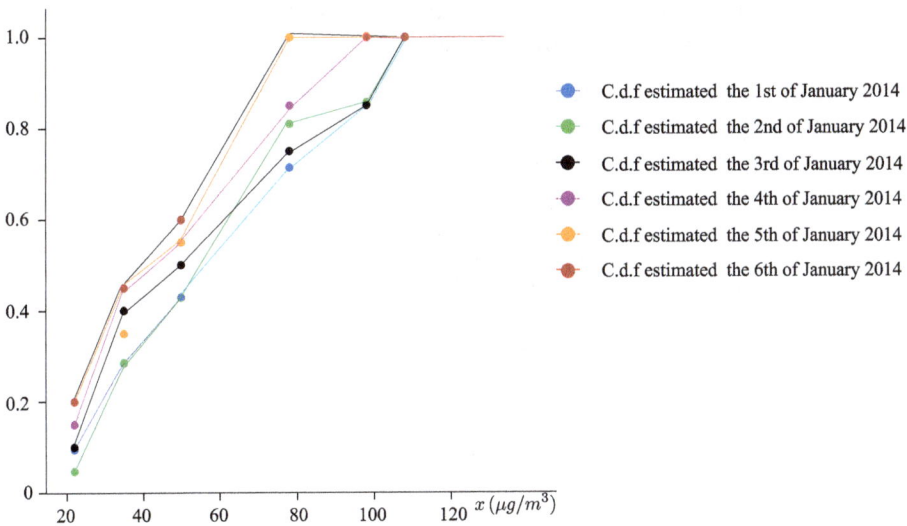

Figure 9. C.d.f.s estimated for PM_{10} daily concentrations ($\mu g/m^3$) at days 1-6 of January 2014

Thresholds	January 2014					
	1st	2nd	3rd	4th	5th	6th
22	0.095	0.047	0.100	0.150	0.200	0.200
35	0.285	0.286	0.400	0.450	0.350	0.450
50	0.429	0.429	0.500	0.550	0.550	0.600
78	0.714	0.810	0.750	0.850	0.900	0.900
98	0.857	0.857	0.850	0.900	0.950	1
108	1	1	1	1	1	1

Table 4. Estimated values for c.d.f. at days 1-6 of January 2014, for fixed thresholds

Moreover, note that it is almost sure that PM_{10} concentrations do not exceed the cutoff 78 $\mu g/m^3$ at days the 5th and 6th (Table 4).

The probability that the variable under study doesn't exceed the threshold fixed by the National Law ($50 \, \mu g/m^3$) is high (equal to 60%) for the last day of interest. In fact, the 6th of January is a non-working day (Epiphany) characterized by low traffic and low industrial emissions.

The local government could use these results in order to carry out environmental policies for the control of high levels of PM_{10}, since it is well known that high concentrations of this pollutant are dangerous for the human health.

Indeed, the estimation of the c.d.f. is a very powerful tool since any action of environmental protection might be adopted in advance by taking into account the actual likelihood of dangerous PM_{10} exceeding. For example, decisions about traffic limitation in high traffic urban area might be supported by the knowledge of the probability that a hazardous pollutant exceeds the level of attention.

8. Conclusions

In this paper, PM_{10} time series analysis, by using geostatistical techniques, has been discussed and the importance of appropriate tools of Geostatistics to study the temporal evolution of this environmental phenomena has been highlighted.

The seasonal behavior of PM_{10} levels has been evaluated through the variogram, that is the basic tool of Geostatistics. Moreover, estimation and prediction problems in the analysis of the time series of this pollutant, characterized by a periodic behavior, have been solved through kriging geostatistical techniques.

The computational aspects have been performed through the use of "KT3D" for the observed values and "KT3DP" for the residuals obtained after removing the periodic component.

Finally, the indicator approach and its capability for assessing the probability that PM_{10} exceeds the specific threshold values have been demonstrated.

The results obtained in this paper by applying geostatistical techniques to analyze PM_{10} time series could be useful to support national policies for environmental and health protection. Governments' activity must be oriented to control that the concentrations of the analyzed pollutant don't exceed specific thresholds according to national or international directives, since it has been demonstrated that particulate matter is dangerous for the human health.

Author details

Claudia Cappello, Sabrina Maggio*, Daniela Pellegrino and Donato Posa

*Address all correspondence to: sabrina.maggio@unisalento.it

University of Salento, Dept. of Management, Economics, Mathematics and Statistics, Italy

References

[1] Arden Pope III C., Dockery D. W. Health effects of fine particulate air pollution: lines that connect. Journal of Air and Waste Management Association 2006; 56 709–742.

[2] Bisgaard S., Khachatryan D. Asymptotic confidence intervals for variograms of stationary time series. Quality and Reliability Engineering International 2010; 26(No. 3) 259–265.

[3] Bloomfield P. Fourier analysis of time series: An Introduction. USA:Wiley & Sons Inc.; 2000.

[4] Box GEP., Jenkins GM. Time series analysis: Forecasting and Control. San Francisco:Holden Day; 1976.

[5] Brockwell P.J., Davis R.A. Time Series: Theory and Methods. New York: Springer; 1987.

[6] Chilés J.P., Delfiner P. Geostatistics: Modeling Spatial Uncertainty. New York: Wiley; 1999.

[7] Cressie, N. (1988). A graphical procedure for determining nonstationarity in time series, Journal of the American Statistical Association 1988; 83(No. 44) 1108–1116.

[8] Cressie, N. (1993). Statistics for Spatial Data, Wiley Series in Probability and Mathematical Statistics, Wiley, New York.

[9] Cressie N., Hawkins D.M. Robust estimation of the variogram. Mathematical Geology 1980;12(2) 115–125.

[10] Cressie N., Grondona M.O. A comparison of variogram estimation with covariogram estimation. In: Mardia K.V. (ed.) The Art of Statistical Science. Chichester: Wiley; 1992.

[11] De Cesare L., Myers D.E., Posa D. FORTRAN 77 programs for space-time modeling. Computers & Geosciences 2002;28(2) 205–212.

[12] De Iaco S., Palma M., Posa, D. Geostatistics and the role of variogram in time series analysis: a critical review. In: Montrone S. & Perchinunno P. (ed.) Statistical Methods for Spatial Planning and Monitoring. Italia: Springer-Verlag; 2013. p47–75.

[13] Deutsch C.V., Journel, A.G. GSLib: Geostatistical Software Library and User's Guide. New York: Oxford University Press; 1998.

[14] Gevers M. On the use of variograms for the prediction of time series. Systems & Control Letters 1985;6(1) 15–21.

[15] Haslett J. On the sample variogram and sample autocovariance for non-stationary time series. Statistician 1997;46(4) 475–485.

[16] Janis M.J., Robeson S.M. Determining the spatial representativeness of air-temperature records using variogram-nugget time series. Physical Geography 2004;25(6) 513–530.

[17] Journel A.G. Nonparametric estimation of spatial distributions. Mathematical Geology 1983;15(3) 445–468.

[18] Journel A.G., Huijbregts C.J. Mining Geostatistics. London: Academic; 1981.

[19] Khachatryan D., Bisgaard S. Some results on the variogram in time series analysis. Quality and Reliability Engineering International 2009;25 947–960.

[20] Little R.J.A., Rubin D.B. Statistical Analysis with Missing Data. New York: Wiley; 2002.

[21] Luceno A. Estimation of missing values in possibly partially nonstationary vector time series. Biometrika 1997;84(2) 495–499.

[22] Matheron G. Principles of Geostatistics. Economic Geology 1963;58(8) 1246–1266.

[23] Matheron G. Les variables régionalisées et leur estimation. Paris: Masson; 1965.

[24] Matheron G. The intrinsic random functions and their applications. Advances in Applied Probability 1973;5(3) 439–468.

[25] Myers D.E. Interpolation with positive definite functions. Science de la Terre 1988;28 251–265.

[26] Myers D.E. Kriging, cokriging, radial basis functions and the role of positive definiteness. Computers & Mathematics with Applications 1992;24(12) 139–148.

[27] Posa D. The indicator formalism in spatial data analysis. Journal of Applied Statistics 1992;19(1) 83–101.

[28] Posa D., Rossi M. Applying stationary and non-stationary kriging. Metron XLVII 1989;(1-4) 295–312.

[29] Posa D., De Iaco S. Geostatistica: teoria e applicazioni. Torino: Giappichelli editore; 2009.

[30] Uysal M. Reconstruction of time series data with missing values. Journal of Applied Sciences 2007;7(6) 922–925.

[31] Weerasinghe S. A missing values imputation method for time series data: an efficient method to investigate the health effects of sulphur dioxide levels. Environmetrics 2010;21(2) 162–172.

## 4

A Non-Homogeneous Markov Chain Model to Study Ozone Exceedances in Mexico City

Eliane R. Rodrigues, Mario H. Tarumoto and Guadalupe Tzintzun

Additional information is available at the end of the chapter

1. Introduction

In many cities around the world, air pollution is among the many environmental problems that affect their population. Among the many known facts about the impact of pollution on human health, we have that for ozone concentration levels above 0.11 parts per million (0.11ppm), the susceptible part of the population (e.g., the elderly, ill, and newborn) staying in that environment for a long period of time, may experience serious health deterioration (see, for example, [1–10]). Therefore, to understand the behaviour of ozone and/or pollutants in general, is a very important issue.

It is possible to find in the literature a vast amount of works that try to answer some of the many issues arising in the study of pollutants' behaviour. Depending on the type of questions that one is trying to answer, different methodologies may be used. Among the many works concentrating on the study of ozone behaviour are, [11–13] using extreme value theory to study the behaviour of the maximum ozone measurements; [14] using time series analysis; [15] using volatility models to study the variability of the weekly average ozone measurements; [13, 16] using homogeneous Poisson processes and [17, 18] using non-homogeneous Poisson models to analyse the probability of having a certain number of ozone exceedances in a time interval of interest; [19] using compound Poisson models to study the occurrence of clusters of ozone exceedances as well as their mean duration time; and [20] using queueing model to study the occurrence of cluster of ozone exceedances as well as their size distribution.

In the environmental area, it is also possible to find works using Markov chains models. Some of them are, [21, 22] where non-homogeneous Markov models are used to study the occurrence of precipitation. We also have [23] where those types of models are used to study tornado activity. In the case of ozone modelling we have, for instance, the works of [24–26] using time homogeneous Markov chains. In those works the interest was in estimating the probability that the ozone measurement would be above (below) a given threshold, conditioned on where it lays in the present and in the past days.

In [24], the order of the Markov chain was estimated using auto-correlation function. Its transition matrix was estimated using the maximum likelihood method (see, for instance, [27, 28], among others). In [25], the order of the chain was also considered an unknown quantity that needed to be estimated. The Bayesian approach (see, for example, [29]) was used to estimate the order as well as the transition probabilities of the chain. In particular, the maximum *à posteriori* method was used. In [26], the estimation of the order of the chain is performed using the Bayesian approach using the so-called trans-dimensional Markov chain Monte Carlo algorithm ([30, 31]). The transition matrix of the chain was obtained through the maximum *à posteriori* method. However, the common denominator of those works is that the Markov chain model used was a time homogeneous one. Since ozone data are not, in general, time homogeneous, the data had to be split into time homogeneous segments and the analysis was made for each segment separately.

Here, the interest also resides in estimating, for instance, the probability that the ozone measurement will be above a given threshold some days into the future, given where it stands today and in the past few days. Although in the present work we also use Markov chain models and the Bayesian approach, the novelty here is that the time-homogeneous assumption is dropped. Here, we consider a non-homogeneous Markov chain model. We assume that the order of the chain as well as its transition probabilities are unknown and need to be estimated. The chosen method of estimation is also the maximum *à posteriori*.

This work is presented as follows. In Section 2 the non-homogeneous Markov chain model is given. Section 3 presents the Bayesian formulation of the model. An application to ozone measurements from Mexico City is given in Section 4. In Section 5 some comments about the methodology and results are made. In an Appendix, before the list of references, we present the code of the programme used to estimate the order and the transition probabilities of the Markov chain.

2. A non-homogeneous Markov chain model

The mathematical model considered here may be described as follows. Let $N > 0$ be a natural number representing the number of years in which measurements were taken. Let $T_i, i = 1, 2, \ldots, N$, be natural numbers representing the amount of observations in each year. Hence, we have that for a given year i, either $T_i = 366$ or $T_i = 365$, depending on whether or not we have a leap year, $i = 1, 2, \ldots, N$.

Let $Z_t^{(i)}$ be the ozone concentration on the tth day of the ith year, $t = 1, 2, \ldots, T_i, i = 1, 2, \ldots, N$. Following [23], we will set $T_i = T = 366, i = 1, 2, \ldots, N$, with the convention that for non leap year, we assign $Z_T^{(i)} = 0$.

Remark. Since, we are taking all years of the same length we will drop the index i from the notation.

Denote by $L > 0$ the environmental threshold we are interested in knowing if the ozone concentration has surpassed or not. Define $\mathbf{Y} = \{Y_t : t \geq 0\}$ by,

$$Y_t = \begin{cases} 0, & \text{if } Z_t < L \\ 1, & \text{if } Z_t \geq L. \end{cases} \tag{1}$$

Hence, Y_t indicates whether or not in the tth day the threshold L was exceeded.

As in [25], we assume that $\mathbf{Y}$ is ruled by a Markov chain of order $K \geq 0$. In contrast with that work, in the present case the Markov chain is a non-homogeneous one. Hence, denote by $X^{(K)} = \{X_t^{(K)} : t = 1, 2, \ldots T\}$, the corresponding non-homogeneous Markov chain of order K. We assume that K has as state space a set $\mathcal{S} = \{0, 1, \ldots, M\}$, for some fixed integer $M \geq 0$, such that, $M \leq T$ with probability one.

Note that, $X^{(K)}$ has as state space the set $S_1^{(K)} = \{(x_1, x_2, \ldots, x_K) \in \{0,1\}^K\}$, with $S_1^{(0)} = S_1^{(1)}$. Also, note that (see [25]), if the set of observed value is $(y_1, y_2, \ldots, y_T)$, then the transition probabilities of $X^{(K)}$ are such that

$$P(X_{t+1}^{(K)} = w \mid X_t^{(K)} = x_t = (y_{t+1}, y_{t+2}, \ldots, y_{t+K})),$$

is different of zero if, and only if, $w = (y_{t+2}, y_{t+3}, \ldots, y_{t+K+1}) \in S_1^{(K)}$, with $0 \leq t \leq T - K$. Therefore, w occurs, if and only if, the observation following $y_{t+1}, y_{t+2}, \ldots, y_{t+K}$, is y_{t+K+1}. This enables us to work with a more treatable state space for $X^{(K)}$, and therefore, to have a better form for the transition matrix.

Hence, as in [25, 32], we consider the transformed state space $S_2^{(K)} = \{0, 1, \ldots, 2^K - 1\}$, which is obtained from $S_1^{(K)}$ by using the transformation $f : S_1^{(K)} \to S_2^{(K)}$, given by, $f(w_1, w_2, \ldots, w_K) = \sum_{l=0}^{K-1} w_{l+1} 2^l$. Let $(x_1, x_2, \ldots, x_K) \leftrightarrow \overline{m}$ indicate that the state $(x_1, x_2, \ldots, x_K) \in S_1^{(K)}$ corresponds to the state $\overline{m} \in S_2^{(K)}$. Hence, the transition probabilities of $X^{(K)}$ may be written as (see, for instance, [25]),

$$P_{\overline{m}j}^{(K)}(t) = P(Y_{t+K+1} = j \mid X_t^{(K)} = (y_{t+1}, y_{t+2}, \ldots, y_{t+K}) \leftrightarrow \overline{m}), \qquad (2)$$

where $\overline{m} \in S_2^{(K)}$, $j \in \{0,1\}$, and $0 \leq t \leq T - K$.

Now, indicate by $Q_{\overline{m}}^{(K)}(t)$, $\overline{m} \in S_2^{(K)}$, $\overline{m} \in S_2^{(K)}$, the probability $P(X_t^{(K)} = \overline{m})$. Hence, when $t = 1$, we have that $Q_{\overline{m}}^{(K)}(1)$, $\overline{m} \in S_2^{(K)}$, is the initial distribution of $X^{(K)}$. When $K = 0$, we have that $P_{\overline{m}j}^{(0)}(t) = Q_j^{(0)}(t)$, $t = 1, 2, \ldots, T$, $j = 0, 1$, $\overline{m} \in S_2^{(0)} = \{0, 1\}$.

Remarks. 1. When $K = 1$, we have that $P_{\overline{m}j}^{(0)}(t)$, $j = 0, 1$, $t = 1, 2, \ldots, T - K$, are the usual one-step transition probabilities. When $K = 0$, the transition probabilities are just the probabilities $Q_{\overline{m}}^{(0)}(t)$, associated to each state $\overline{m} \in S_2^{(0)} = \{0, 1\}$, with $t = 1, 2, \ldots, T$.

2. Unless otherwise stated, from now on, we are going to use the state space $S_2^{(K)}$ and the corresponding transition probabilities.

3. $\mathbf{Y}$ is going to represent our observed data.

In addition to estimating the order K of the Markov chain, we will also estimate its transition probabilities $P_{\overline{m}j}^{(K)}(t)$ as well as the probabilities $Q_{\overline{m}}^{(K)}(t)$, $j \in \{0, 1\}$, $\overline{m} \in S_2^{(K)}$, for each t. We indicate by $P^{(K)}(t) = \left(P^{(K)}_{\overline{m}j}(t)\right)_{j \in \{0,1\}, \overline{m} \in S_2^{(K)}}$, the transition matrix at time t. Note that, if $K = 0$, then $P^{(0)}_{\overline{m}j}(t) = Q_j^{(0)}(t), j \in \{0,1\}, \overline{m} \in S_2^{(0)}, t = 1, 2, \dots, T$.

3. A Bayesian estimation of the parameters of the model

There are many ways of estimating the order and the transition matrix of a non-homogeneous Markov chain. One way of estimating the order is via the auto-correlation function associated to the chain throughout the years. Another way is to use the Bayesian approach. When it comes to estimating the transition probabilities we have, for instance, the maximum likelihood method ([33]) and the empirical estimator ([34]) which are essentially the same. In the present work, we will use the Bayesian approach (see, for instance, [29, 35]) to estimate the order and the transition probabilities. In particular, we are going to adopt the maximum *à posteriori* approach. Inference will be performed using the information provided by the so-called posterior distribution of the parameters. The posterior distribution of a vector of parameters $\boldsymbol{\theta}$ given the observed data $\mathbf{D}$, indicated by $P(\boldsymbol{\theta} \mid \mathbf{D})$, is such that $P(\boldsymbol{\theta} \mid \mathbf{D}) \propto L(\mathbf{D} \mid \boldsymbol{\theta}) P(\boldsymbol{\theta})$, where $L(\mathbf{D} \mid \boldsymbol{\theta})$ is the likelihood function of the model, and $P(\boldsymbol{\theta})$ is the prior distribution of the vector $\boldsymbol{\theta}$.

In the present case, we have that the vector of parameter is $\boldsymbol{\theta} = (K, Q^{(K)}(1), P^{(K)}(t), t = 1, 2, \dots, T - K)$. If $K = 0$, the range of t is $\{1, 2, \dots, T\}$. The vector $\boldsymbol{\theta}$ belongs to the following sample space

$$\Theta = \bigcup_{K=0}^{M} \left(\{K\} \times \Delta_2^{2^K} \times \Delta_2^{(T-K) 2^K}\right)$$

where $\Delta_2 = \{(x_1, x_2) \in \mathbb{R}^2 : x_i \geq 0, i = 1, 2, x_1 + x_2 = 1\}$ is the one dimensional simplex. (Note that if we have $K = 0$, then the parametric space reduces to $\Theta = \Delta_2^T$.) In the present case we have $\mathbf{D} = \mathbf{Y}$

Let $(x_1, x_2, \dots, x_K)$ be such that $Y_t = x_1$. Indicate by $n^{(K)}_{\overline{m}i}(t)$ the number of years in which the vector $(x_1, x_2, \dots, x_K)$ corresponding to a state $\overline{m} \in S_2^{(K)}$ is followed by the observation i, $i = 0, 1$. Also define $n^{(0)}_{\overline{m}}(t), \overline{m} \in S_2^{(0)} = \{0, 1\}$, as the number of years in which we have the observation $\overline{m}$ at time t, $t = 1, 2, \dots, T$. Additionally, let $n^{(K)}_{\overline{m}}$ indicate the number of years in which the state corresponding to the initial K days is equivalent to the value $\overline{m} \in S_2^{(K)}$, $K \geq 0$. In the case of $K = 0$, we have $n^{(0)}_{\overline{m}} = n^{(1)}_{\overline{m}}$, and $\overline{m} \in S_2^{(0)} = S_2^{(1)} = \{0, 1\}$.

Therefore, since a Markovian model is assumed, the likelihood function is given by (see, for instance, [23, 33])

$$L(\mathbf{Y} \mid \boldsymbol{\theta}) \propto \left(\prod_{\overline{m} \in S_2^{(K)}} \left[Q^{(K)}_{\overline{m}}(1)\right]^{n^{(K)}_{\overline{m}}}\right) \left(\prod_{t=1}^{T-K} \left[P^{(K)}_{\overline{m}0}(t)\right]^{n^{(K)}_{\overline{m}0}(t)} \left[1 - P^{(K)}_{\overline{m}0}(t)\right]^{n^{(K)}_{\overline{m}1}(t)}\right). \qquad (3)$$

Note that when $K = 0$, the expression (3) simplifies to

$$L(\mathbf{Y} \mid \boldsymbol{\theta}) \propto \prod_{t=1}^{T} \prod_{m=0}^{1} \left[Q_{\overline{m}}^{(0)}(t) \right]^{n_{\overline{m}}^{(0)}(t)},$$

where $Q_{\overline{m}}^{(0)}(t) = P(X_t^{(0)} = \overline{m}) = P(Y_t = \overline{m})$, $t = 1, 2, \ldots, T$, $\overline{m} \in S_2^{(0)} = \{0, 1\}$.

The prior distribution of the vector of parameters is given as follows. We assume a prior independence of $P^{(K)}(t)$ as functions of t. Also, since the forms of $P^{(K)}(t)$ and $Q^{(K)}(1)$ depend on the value of K, we have that for $\boldsymbol{\theta} = (K, Q^{(K)}(1), P^{(K)}(t), t = 1, 2, \ldots, T - K)$,

$$P(\boldsymbol{\theta}) = P(Q^{(K)}(1) \mid K) \left[\prod_{t=1}^{T-K} P\left(P^{(K)}(t) \mid K \right) \right] P(K),$$

where $P(Q^{(K)}(1) \mid K)$ and $P\left(P^{(K)}(t) \mid K \right)$ are the prior distributions of the initial distribution $Q^{(K)}(1)$ and of the transition matrix $P^{(K)}(t)$ given the order of the chain, respectively, and $P(K)$ the prior distribution of the order K.

Remark. When we have $K = 0$, the vector of parameters is $\boldsymbol{\theta}' = (Q_{\overline{m}}^{(0)}(t); t = 1, 2, \ldots, T)$, whose prior distribution is

$$P(\boldsymbol{\theta}') = \left[\prod_{t=1}^{T} P(Q^{(0)}(t)) \right],$$

where $P(Q^{(0)}(t))$ is the prior distribution of the probability vector $Q^{(0)}(t) = (Q_{\overline{m}}^{(0)}(t), \overline{m} = 0, 1)$, $t = 1, 2, \ldots, T$.

Given the nature of transition matrices, we are going to assume that rows are independent. We also assume that, given the order K of the chain, each row of the transition matrix $P^{(K)}(t)$ will have as prior distribution a Dirichlet distribution with appropriate hyperparameters. Therefore, given that $K = k$, row $(P_{\overline{m}0}^{(k)}(t), P_{\overline{m}1}^{(k)}(t))$ has as prior distribution a Dirichlet$(\alpha_{\overline{m}0}^{(K)}(t), \alpha_{\overline{m}1}^{(K)}(t))$, $t = 1, 2, \ldots, T$; i.e.,

$$P\left(P^{(K)}(t) \mid K \right) = \prod_{\overline{m} \in S_2^{(K)}} \left(\frac{\Gamma(\alpha_{\overline{m}0}^{(K)}(t) + \alpha_{\overline{m}1}^{(K)}(t))}{\Gamma(\alpha_{\overline{m}0}^{(K)}(t)) \, \Gamma(\alpha_{\overline{m}1}^{(K)}(t))} \left\{ \left[P_{\overline{m}0}^{(K)}(t) \right]^{\alpha_{\overline{m}0}^{(K)}(t)-1} \left[P_{\overline{m}1}^{(K)}(t) \right]^{\alpha_{\overline{m}1}^{(K)}(t)-1} \right\} \right)$$

for t in the appropriate range. In the case of initial distribution $Q^{(K)}(1)$, we also have a Dirichlet prior distribution, but now with hyperparameters $(\alpha_{\overline{m}}^{(K)}; \overline{m} \in S_2^{(K)})$. Therefore,

$$P\left(Q^{(K)}(1) \mid K \right) = \frac{\Gamma(\sum_{\overline{m} \in S_2^{(K)}} \alpha_{\overline{m}}^{(K)})}{\prod_{\overline{m} \in S_2^{(K)}} \Gamma(\alpha_{\overline{m}}^{(K)})} \prod_{\overline{m} \in S_2^{(K)}} \left[Q_{\overline{m}}^{(K)}(1) \right]^{\alpha_{\overline{m}}^{(K)}-1}.$$

If $K = 0$, then $Q^{(0)}(t)$ has as prior distribution a Dirichlet distribution with hyperparameters $(\alpha_{\overline{m}}^{(0)}(t); t = 1, 2, \ldots, T; \overline{m} \in S_2^{(0)} = \{0, 1\})$. Therefore, for any given t,

$$P\left(Q^{(0)}(t) \mid K = 0\right) = \frac{\Gamma(\sum_{\overline{m}=0}^{1} \alpha_{\overline{m}}^{(0)}(t))}{\prod_{\overline{m}=0}^{1} \Gamma(\alpha_{\overline{m}}^{(0)}(t))} \prod_{\overline{m}=0}^{1} \left[Q_{\overline{m}}^{(0)}(t)\right]^{\alpha_{\overline{m}}^{(0)}(t)-1}.$$

We assume that K has as prior distribution a truncated Poisson distribution defined on the set $\mathcal{S}$ with rate $\lambda > 0$; i.e.,

$$P(K) = \frac{\lambda^K}{K!} I_{\mathcal{S}}(K),$$

where $I_A(x) = 1$, if $x \in A$ and is zero otherwise.

Therefore, we have from [25, 32, 36], that the conditional posterior distribution of $P^{(K)}(t)$ given K, is

$$P\left(P^{(K)}(t) \mid K, \mathbf{Y}\right) \propto \prod_t \left\{ \prod_{\overline{m} \in S_2^{(K)}} \left(\left[P_{\overline{m}0}^{(K)}(t)\right]^{n_{\overline{m}0}^{(K)}(t)+\alpha_{\overline{m}0}^{(K)}(t)-1} \left[P_{\overline{m}1}^{(K)}(t)\right]^{n_{\overline{m}1}^{(K)}(t)+\alpha_{\overline{m}1}^{(K)}(t)-1} \right) \right\}.$$

Hence, $P\left(P^{(K)}(t) \mid K, \mathbf{Y}\right)$ is proportional to the product of Dirichlet distributions with hyperparameters $(n_{\overline{m}0}^{(K)}(t) + \alpha_{\overline{m}0}^{(K)}(t), n_{\overline{m}1}^{(K)}(t) + \alpha_{\overline{m}1}^{(K)}(t))$. The mode of each Dirichlet distribution is known and is given by (see [37]),

$$P_{\overline{m}i}^{(K)}(t) = \frac{n_{\overline{m}i}^{(K)}(t) + \alpha_{\overline{m}i}^{(K)}(t) - 1}{\sum_{j=0}^{1} \left[n_{\overline{m}j}^{(K)}(t) + \alpha_{\overline{m}j}^{(K)}(t) - 1\right]}, \quad i = 0, 1; \overline{m} \in S_2^{(K)}; K \in \mathcal{S}; t = 1, 2, \ldots, T - K. \quad (4)$$

Additionally, the posterior distribution of the initial distribution $Q^{(K)}(1)$ given K is

$$P(Q^{(K)}(1) \mid K, \mathbf{Y}) \propto \prod_{\overline{m} \in S_2^{(K)}} \left[Q_{\overline{m}}^{(K)}(1)\right]^{\alpha_{\overline{m}}^{(K)}+n_{\overline{m}}^{(K)}-1}.$$

Therefore, $P(Q^{(K)}(1) \mid K, \mathbf{Y})$ is proportional to a Dirichlet distribution with hyperparameters $(\alpha_{\overline{m}}^{(K)} + n_{\overline{m}}^{(K)}; \overline{m} \in S_2^{(K)})$. Hence, as in the case of the posterior distribution of $P^{(K)}(t)$, the mode of $P(Q^{(K)}(1) \mid K, \mathbf{Y})$ is,

$$Q_{\overline{m}}^{(K)}(1) = \frac{n_{\overline{m}}^{(K)} + \alpha_{\overline{m}}^{(K)} - 1}{\sum_{\overline{m}' \in S_2^{(K)}} \left[n_{\overline{m}'}^{(K)} + \alpha_{\overline{m}'}^{(K)} - 1\right]}, \quad \overline{m} \in S_2^{(K)}; K \in \mathcal{S}. \quad (5)$$

When $K = 0$, we have

$$P(Q^{(0)}(t) \mid K = 0, \mathbf{Y}) \propto \left(\prod_{\overline{m}=0}^{1} \left[Q_{\overline{m}}^{(0)}(t) \right]^{n_{\overline{m}}^{(0)}(t) + \alpha_{\overline{m}}^{(0)}(t) - 1} \right), \quad t = 1, 2, \ldots, T.$$

Therefore, for each $t = 1, 2, \ldots, T$,

$$Q_{\overline{m}}^{(0)}(t) = \frac{n_{\overline{m}}^{(0)}(t) + \alpha_{\overline{m}}^{(0)}(t) - 1}{\sum_{\overline{m}'=0}^{1} \left[n_{\overline{m}'}^{(0)}(t) + \alpha_{\overline{m}'}^{(0)}(t) - 1 \right]}, \quad \overline{m} = 0, 1. \tag{6}$$

Furthermore, we also have, from [25], that

$$L(\mathbf{Y} \mid K) \propto \frac{\Gamma\left(\sum_{\overline{m} \in S_2^{(K)}} \alpha_{\overline{m}}^{(K)} \right)}{\Gamma\left(\sum_{\overline{m} \in S_2^{(K)}} \left[\alpha_{\overline{m}}^{(K)} + n_{\overline{m}}^{(K)} \right] \right)} \left(\prod_{\overline{m} \in S_2^{(K)}} \frac{\Gamma\left(n_{\overline{m}}^{(K)} + \alpha_{\overline{m}}^{(K)} \right)}{\Gamma\left(\alpha_{\overline{m}}^{(K)} \right)} \right)$$

$$\prod_{t=1}^{T-K} \left\{ \prod_{\overline{m} \in S_2^{(K)}} \left(\frac{\Gamma[\alpha_{\overline{m}0}^{(K)}(t) + \alpha_{\overline{m}1}^{(K)}(t)]}{\Gamma(\sum_{j=0}^{1} [n_{\overline{m}j}^{(K)}(t) + \alpha_{\overline{m}j}^{(K)}(t)])} \prod_{j=0}^{1} \frac{\Gamma(n_{\overline{m}j}^{(K)}(t) + \alpha_{\overline{m}j}^{(K)}(t))}{\Gamma(\alpha_{\overline{m}j}^{(K)}(t))} \right) \right\},$$

with the appropriate adaptation for the case of $K = 0$. Hence, the posterior distribution of the order K is

$$P(K \mid \mathbf{Y}) = \frac{1}{c} L(\mathbf{Y} \mid K) \frac{\lambda^K}{K!} \tag{7}$$

where $c = \sum_{k \in \mathcal{S}} L(\mathbf{Y} \mid K = k) \left(\lambda^k / k! \right)$ is the normalising constant.

Therefore, in order to obtain the probability of interest, we just have to use (7) to estimate the value of K that maximises that posterior probability, and then use (4), (5), and/or (6) (depending on the case), in order to calculate the corresponding transition matrix and initial distribution, respectively.

The hyperparameters appearing in the prior distribution will be considered known and will be specified later.

4. Application to ozone data from the monitoring network of Mexico City

In this section we apply the model to the Mexico City's ozone measurements. The data used consist of twenty two years of the daily maximum ozone measurements (from 01 January 1990 to 31 December 2011) provided by the monitoring network of the Metropolitan Area of Mexico City. The Metropolitan Area is divided into five regions, namely, Northeast (NE), Northwest (NW), Centre (CE), Southeast (SE), and Southwest (SW). The monitoring stations are placed throughout the city. Measurements in each monitoring station are obtained minute by minute and the averaged hourly result is reported at each station. The daily maximum measurement for a given region is the maximum over all the maximum averaged values recorded hourly during a 24-hour period by each station placed in the region. Since emergency alerts in Mexico City are declared regionally, we will analyse each region separately.

The Mexican ozone standard considers the threshold 0.11ppm (see [38]). Hence, we will take that value as one of our thresholds. Additionally, for comparison purpose, we will also take the threshold values 0.15ppm and 0.17ppm. One of the reasons for choosing these latter values is that we would like to know what would happen if the threshold for declaring emergency alerts in Mexico City was lowered to 0.17ppm. The reason for choosing the threshold 0.15ppm is because it is an intermediate value between the Mexican standard and 0.17ppm.

During the observational period considered here, we have that the mean of the daily observed measurements were 0.12, 0.098, 0.13, 0.12, and 0.14, in regions NE. NW, CE, SE, and SW, respectively, with corresponding standard deviations of 0.06, 0.04, 0.06, 0.05, and 0.06, for those same regions. The threshold 0.11ppm was either reached or exceeded in 4280, 3139, 4921, 4921, and 5711 days in regions NE, NW, CE, SE, and SW, respectively. In those same regions, the threshold 0.15ppm was reached or exceeded in 2460, 963, 2819, 2299, and 3594 days, and the numbers in the case of the threshold 0.17ppm are, 1769, 479, 1896, 1419, and 2660, respectively.

Even though it is a general belief that ozone measurements depend on the measurements of only a few days in the past, we are taking $M = 16$ when we consider the threshold values 0.15ppm and 0.17ppm. We have decided to do that because in previous works the order for homogeneous segments could have higher order. In the case of $L = 0.11$ppm, in some cases, larger values of M were needed. Hence, we also take $M = 16$, in the case of region NW, and we take $M = 18$ in the case of regions CE, NE, SE, and SW. In order to account also for the possibility of low order, we take $\lambda = 1$ in the prior distribution of K.

The hyperparameters of the Dirichlet prior distributions are assigned as in [25]. Therefore, the values of $\alpha_{\overline{mi}}^{(K)}(t)$, $\alpha_{\overline{m}}^{(0)}(t)$, and $\alpha_{\overline{m}}^{(K)}$ will belong to the set $\{3,4,5,6,7,8\}$. Hence, assign $\alpha_{\overline{mi}}^{(K)}(t) = 8$ for the coordinate corresponding to the $\max\{n_{\overline{m0}}^{(K)}(t), n_{\overline{m1}}^{(K)}(t)\}$. Depending on the difference $\max\{n_{\overline{m0}}^{(K)}(t), n_{\overline{m1}}^{(K)}(t)\} - \min\{n_{\overline{m0}}^{(K)}(t), n_{\overline{m1}}^{(K)}(t)\}$, an integer value in $\{3,4,5,6,7\}$ is assigned to the hyperparameter corresponding to $\min\{n_{\overline{m0}}^{(K)}(t), n_{\overline{m1}}^{(K)}(t)\}$. If we have $n_{\overline{mi}}^{(K)}(t) = 0$, then the value 3 is automatically assigned to the corresponding $\alpha_{\overline{mi}}^{(K)}(t)$. Similar procedure is applied in the cases of $\alpha_{\overline{m}}^{(0)}(t)$ and $\alpha_{\overline{m}}^{(K)}$.

Table 1 gives the values of $P(K \mid Y)$. Even though, S includes the values 0, 1, 2, and 3, since the posterior probabilities at those points are of order 10^{-8} and below, we have omitted those values of K. We use the symbol "-" to indicate that the specific value of K either was not considered in the corresponding region or the probability associated to it was small compared to the values shown.

Looking at Table 1 we may see that, if we consider the threshold $L = 0.11$ppm, then the selected order of the chain is K equal to 16 in the case of region NE, equal to 12 for region NW, and equal to 17 for regions CE and SE. When we consider region SW, the value of K is

	NE			NW			CE			SE			SW		
	0.11	0.15	0.17	0.11	0.15	0.17	0.11	0.15	0.17	0.11	0.15	0.17	0.11	0.15	0.17
$K = 4$	–	–	–	–	–	$< 10^{-7}$	–	–	–	–	–	–	–	–	–
$K = 5$	–	–	–	–	0.33	1	–	–	–	–	–	0.641	–	–	–
$K = 6$	–	–	0.02	–	0.67	$< 10^{-16}$	–	–	–	–	–	0.359	–	–	–
$K = 7$	–	–	0.93	–	–	–	–	–	0.33	–	–	–	–	–	–
$K = 8$	–	0.007	0.05	–	–	–	–	–	0.67	–	0.09	–	–	–	0.024
$K = 9$	–	0.173	–	–	–	–	–	–	–	–	0.9	–	–	–	0.635
$K = 10$	–	0.67	–	–	–	–	–	0.05	–	–	0.01	–	–	–	0.383
$K = 11$	–	0.15	–	0.47	–	–	–	0.68	–	–	–	–	–	–	0.003
$K = 12$	–	–	–	0.53	–	–	–	0.26	–	–	–	–	–	0.008	–
$K = 13$	–	–	–	–	–	–	–	–	–	–	–	–	–	0.095	–
$K = 14$	–	–	–	–	–	–	–	–	–	–	–	–	–	0.792	–
$K = 15$	–	–	–	–	–	–	–	–	–	–	–	0.012	–	0.105	–
$K = 16$	0. 66	–	–	–	–	–	0.01	–	–	0.004	–	–	–	–	–
$K = 17$	0.07	–	–	–	–	–	0.99	–	–	0.984	–	–	–	–	–
$K = 18$	0.27	–	–	–	–	–	–	–	–	–	–	–	–	–	–

Table 1. Posterior distribution of the order of the chain for all regions and threshold considered. The symbol "-" is used to indicate that the specific value of K either was not considered in the corresponding region or the probability associated to it was small compared to the values shown.

either larger than or equal to 18 with probability one. If we take into account the threshold $L = 0.15$ppm, then, also by looking at Table 1, we have that the chosen orders are $10, 6, 11, 9,$ and 14, in the cases of regions NE, NW, CE, SE, and SW, respectively. When we consider the threshold $L = 0.17$ppm, then the selected orders are $7, 8,$ and 9, for regions NE, CE, and SW, respectively. In the cases of regions NW and SE, the estimated order is 5. Therefore, using this information and (4), the corresponding transition and initial probabilities may then be calculated.

As an example, consider the case of region CE and the threshold 0.17ppm. In that case, we have that the order of the chain is $K = 8$. Therefore, $S_2^{(K)} = \{0, 1, \ldots, 255\}$. In Table 2, we have the approximated estimated values of the initial distribution $Q_{\overline{m}}^{(K)}(1)$, and of the transition probabilities $P_{\overline{m}0}^{(K)}(t)$, $t = 1, 2$. (We have truncated the values and the total sum is approximately one.) We use the notation $\overline{m}' - \overline{m}''$, to indicate that for all values of $\overline{m}$ in $\{\overline{m}', \overline{m}' + 1, \ldots, \overline{m}''\}$, the estimated probabilities are equal to the values shown.

Looking at Table 2, it is possible to see that the highest initial probability is that associated to the state $\overline{0}$, i.e., the first eight days of the year form a string of zeros, meaning that the concentration levels are below 0.17ppm. Additionally, once you have the information that the ozone concentration levels on the first eight days are below 0.17ppm, then the highest transition probability is also associated to the transition to zeros, i.e., the two days following the eight initial days with concentration below 0.17ppm are more likely to present lower concentration levels as well.

In order to illustrate the type of information that may be obtained using the methodology considered here, take the case of the year 2012 and region CE. Suppose we want to calculate the probability that during the first nine days of January we have that the ozone concentration is below 0.17ppm from the first eight days, and it is above it on the ninth. Therefore, we want to know the probability that $(0, 0, 0, 0, 0, 0, 0, 0)$ is followed by one. Hence, we want the probability of having the following sequence of zeros and ones: $0, 0, 0, 0, 0, 0, 0, 0, 1$. Therefore,

$$P((0,0,0,0,0,0,0,0,1)) = P_{\overline{0}1}^{(8)}(1) \times Q_{\overline{0}}^{(8)}(1) = 0.238 \times 0.0327 \approx 0.008.$$

$\overline{m}$	$Q_{\overline{m}}^{(8)}(1)$		$\overline{m}$	$P_{\overline{m}0}^{(8)}(1)$	$P_{\overline{m}0}^{(8)}(2)$
$\overline{0}$	0.0327		$\overline{0}$	0.762	0.889
$\overline{1}-\overline{15}$	0.0036		$\overline{1}-\overline{11}$	0.5	0.5
$\overline{16}$	0.0073		$\overline{12}$	0.5	0.286
$\overline{17}-\overline{23}$	0.0036		$\overline{13}-\overline{62}$	0.5	0.5
$\overline{24}$	0.0073		$\overline{63}$	0.286	0.8
$\overline{25}-\overline{27}$	0.0036		$\overline{64}-\overline{127}$	0.5	0.5
$\overline{28}$	0.0073		$\overline{128}$	0.5	0.444
$\overline{29}-\overline{31}$	0.0036		$\overline{129}-\overline{158}$	0.5	0.5
$\overline{32}$	0.0073		$\overline{159}$	0.5	0.286
$\overline{33}-\overline{62}$	0.0036		$\overline{160}-\overline{247}$	0.5	0.5
$\overline{63}$	0.0073		$\overline{248}$	0.286	0.5
$\overline{63}-\overline{125}$	0.0036		$\overline{249}$	0.5	0.5
$\overline{126}$	0.0073		$\overline{250}$	0.286	0.5
$\overline{127}-\overline{191}$	0.0036		$\overline{251}$	0.5	0.6
$\overline{192}$	0.0073		$\overline{252}-\overline{253}$	0.5	0.286
$\overline{193}-\overline{241}$	0.0036		$\overline{254}-\overline{255}$	0.5	0.5
$\overline{242}$	0.0073		$-$	$-$	$-$
$\overline{243}$	0.0036		$-$	$-$	$-$
$\overline{244}$	0.0073		$-$	$-$	$-$
$\overline{245}-\overline{247}$	0.0036		$-$	$-$	$-$
$\overline{248}$	0.0073		$-$	$-$	$-$
$\overline{249}$	0.0036		$-$	$-$	$-$
$\overline{250}$	0.0073		$-$	$-$	$-$
$\overline{251}-\overline{255}$	0.0036		$-$	$-$	$-$

Table 2. Transition probabilities $P_{\overline{m}0}^{(8)}(1)$ and $P_{\overline{m}0}^{(8)}(2)$ as well as the initial probabilities $Q_{\overline{m}}^{(8)}(1)$, for all values of $\overline{m} \in S_2^{(K)}$ in the case of region CE and threshold 0.17ppm. The notation $\overline{m}' - \overline{m}''$ is used to indicate that for all values of $\overline{m}$ in $\{\overline{m}', \overline{m}'+1, \ldots, \overline{m}''\}$, the estimated probabilities are equal to the values shown.

In order to obtain the values of the probabilities of interest, recall that $P_{\overline{m}0}^{(K)}(t) = 1 - P_{\overline{m}1}^{(K)}(t)$, $\overline{m} \in S_2^{(K)}$, $t = 1, 2, \ldots, T - K$. Therefore, looking at Table 2, we have that $P_{\overline{01}}^{(8)}(1)$ is one minus the value on the column corresponding to $P_{\overline{m}0}^{(K)}(1)$ with $\overline{m} = \overline{0}$. Similar comment is valid in the case of $Q_{\overline{m}}^{(8)}(1)$, $\overline{m} \in S_2^{(K)}$.

Suppose now that we want to know the probability of having (0, 0, 0, 0, 0, 0, 0, 0) followed by (1, 1). Hence, we want to know what the probability that (0, 0, 0, 0, 0, 0, 0, 0) is followed by one and that (0, 0, 0, 0, 0, 0, 0, 1) is followed by one. Therefore, we need to calculate

$$P((0,0,0,0,0,0,0,0,1,1)) = P_{\overline{255}1}^{(8)}(2) \times P_{\overline{01}}^{(8)}(1) \times Q_{\overline{0}}^{(8)}(1) = 0.5 \times 0.238 \times 0.0327 \approx 0.0004.$$

Proceeding in this way we may calculate the probability of having any string of states at any time of the year.

If we compare to the actual measurements in the year 2012, then we have that in the first ten days, the sequence **Y**, in the case of region CE, has the configuration 0, 0, 0, 0, 0, 0, 0, 0, 0, 0. In fact, the estimated probability of that sequence of zeros and ones is $0.5 \times 0.762 \times 0.0327 \approx 0.0125$ which is three times higher than the probability of having (0, 0, 0, 0, 0, 0, 0, 0) followed by (1, 1). If we consider also the year 2013, the results are similar. Hence, the methodology used here can produce estimated values that may describe well the behaviour of the data.

5. Conclusion

In this work we have considered a non-homogeneous Markov chain model to study the ozone's behaviour in Mexico City. The interest resides in estimating the probability that the ozone level will be above (below) a certain threshold given that it is either above or below it in the present and in the recent past. Due to the nature of the questions asked here, a natural way of trying to answer them is to use Markov chain models. However, due to the non-homogeneity of the data, a non-homogeneous version of the chain is used.

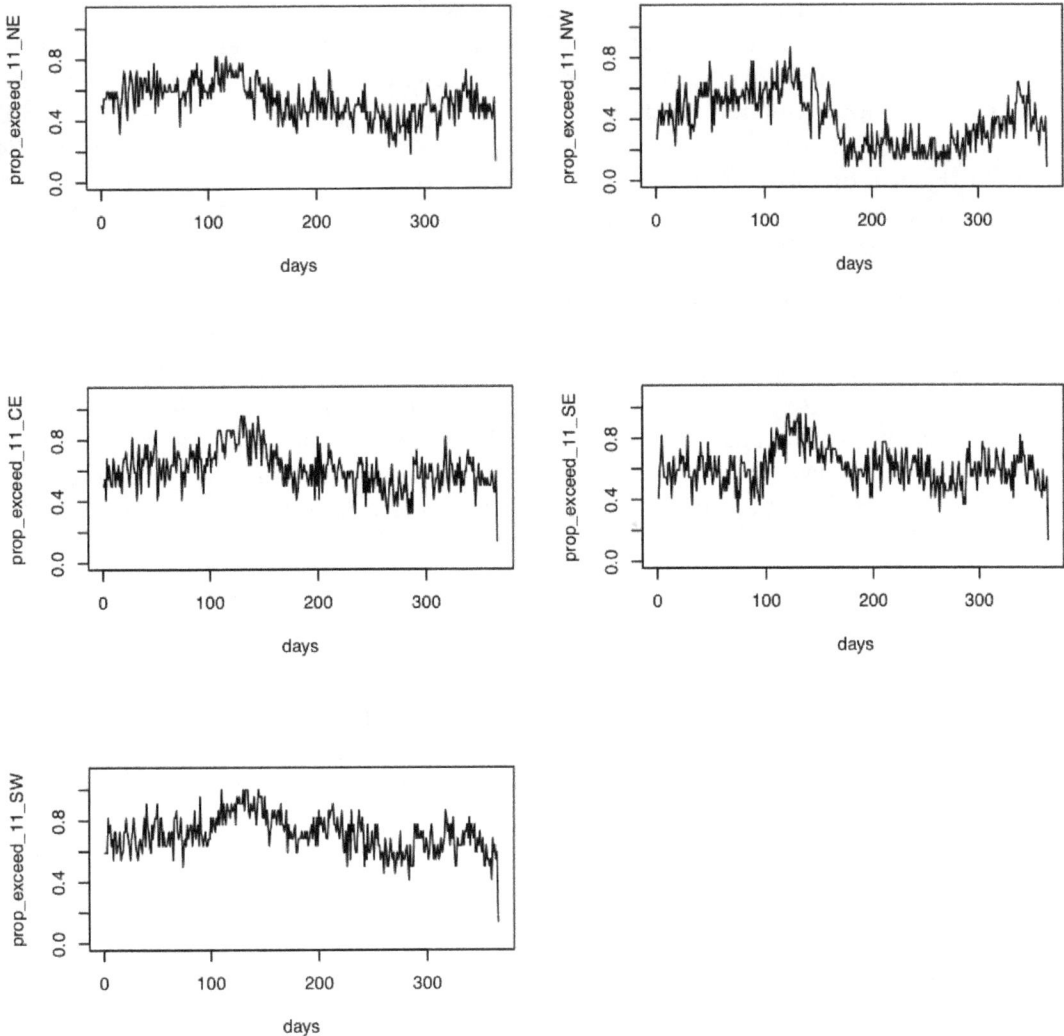

Figure 1. Proportion of years in which, for a given day, the threshold 0.11ppm was exceeded by the ozone concentration.

Using the Bayesian approach a maximum *à posteriori* estimation of the order of the matrix as well as its transition matrix and initial distribution was made. The results have shown that higher order should be considered for the chain. One explanation for that could be the way the empirical probability of having an exceedance in a given day behaves. That can be seen in Figure 1 where, as an illustration, the proportion of exceedances of the threshold 0.11ppm is presented for each region. The values correspond to the proportion of years in which in a fixed day the threshold was exceeded. As we vary the days, we have the behaviour throughout the 366 days. In that figure we have in the horizontal axis the days and in the vertical axis we have the values of $prop(t) = (1/N) \sum_{i=1}^{N} Y_t^{(i)}$, which represents the proportion of years with an exceedance in the tth day, $t = 1, 2, \ldots, T$. The notation $Y_t^{(i)}$ is used to indicate the variable Y_t defined in (1) on the ith year.

The plots in Figure 1 reflect well the fact that in region SW, in most of the days of the year, there are exceedances of the threshold 0.11. We may also see the influence of the seasons of the year. The hill between days 100 and 200 appearing in every plot, corresponds to measurements taken between April and June. Higher values occur during the days corresponding to approximately mid April to mid May. Those months are in the middle of Spring. During this season it does rain much in Mexico City. Additionally, there is a lot of sunlight. Hence, the ozone concentration is bound to be high, and as a consequence, the proportion of years in which exceedances occur at that period is large. The values decrease when the raining season starts (around the beginning of June).

If we consider the threshold values 0.15ppm and 0.17ppm, the behaviour of the proportion of years where in a given day exceedances occurred is similar to the case of 0.11ppm. The difference is that the values of the proportions are smaller. It is possible to see that the proportion of exceedances may vary according to the seasons of the year, and that, within a given season, changes are, in general, not drastic. Therefore, it is possible that measurements from more than a few days may have an influence on the behaviour of future measurements, and with that, make the estimation method considered here to produce high values for the order of the chain.

Acknowledgements

The authors thank the Editor for sending comments that helped to improve the presentation of the results. The authors also thank Peter Guttorp for providing a copy of his works related to applications of non-homogeneous Markov chains. ERR and MHT were partially funded by the project PAPIIT-IN102713-3 of the Dirección General de Apoyo al Personal Académico de la Universidad Nacional Autónoma de México (DGAPA-UNAM), Mexico. ERR also received funds from a sabbatical year grant from DGAPA-UNAM. ERR is grateful to the Departments of Statistics of the Universidade Estadual Paulista "Júlio de Mesquita Filho" – Campus Presidente Prudente, Brazil, and of the University of Oxford, UK, where parts of this work were developed, for all the support and hospitality received during her stay at those departments.

Appendix

In this Appendix we present the code of the programme in R used to estimate the order of the non-homogeneous Markov chain as well as its transition probabilities. The code it not optimal and can be highly improved, but in its present form it provides elements for estimating the necessary quantities.

```
# ESTIMATING THE ORDER OF THE CHAIN
ozonio_sw=read.table('data.txt', header=T)
attach(ozonio_sw)
anos=NCOL(ozonio_sw)
dias=NROW(ozonio_sw)
# assigning the value of M
M = 15 # for instance
# initialisation of the matrices to store the values of the likelihood
# for each order and day
term_init_like <- matrix(0, M+1, 1)
term_general_like <- matrix(0, dias, M+1)
#######
# Case K = 0
#######
# counting the numbers of ones and zeros in each row
count_init_0 <- matrix(0, dias, 2)
for(i in 1:dias){
for(j in 1:anos){
if(ozonio_sw[i,j] == 0){count_init_0[i,1] = count_init_0[i,1] + 1} }
count_init_0[i,2] = anos - count_init_0[i,1]}
# assigning the values of alpha
alpha_init_0 <- matrix(0,dias,2)
for(i in 1:dias){
for(j in 1:2){if(count_init_0[i,j] == 0){alpha_init_0[i,j] = 3} }
if ((count_init_0[i,1] == min(count_init_0[i,])) & (count_init_0[i,1] != 0)){
if(alpha_init_0[i,1] == 0){alpha_init_0[i,1] = 4
if(alpha_init_0[i,2] == 0){alpha_init_0[i,2] = 8}} }
if ((count_init_0[i,1] == max(count_init_0[i,])) & (count_init_0[i,2] != 0)){
if(alpha_init_0[i,1] == 0){alpha_init_0[i,1] = 8
if(alpha_init_0[i,2] == 0){alpha_init_0[i,2] = 4}} }
if ((count_init_0[i,1] == min(count_init_0[i,])) & (count_init_0[i,1] == 0)){
if(alpha_init_0[i,2] == 0){alpha_init_0[i,2] = 8} }
if ((count_init_0[i,1] == max(count_init_0[i,])) & (count_init_0[i,2] == 0)){
if(alpha_init_0[i,1] == 0){alpha_init_0[i,1] = 8} }
}
# calculating the value of the likelihood L(Y | K = 0)
prod_1_k0 <- matrix(0,dias,1)
prod_2_k0 <- matrix(0,dias,1)
prod_k0 <- matrix(0,dias,1)
for(i in 1:dias){
for(j in 1:2){
prod_1_k0[i] = (gamma(count_init_0[i,j]+alpha_init_0[i,j])/gamma(alpha_init_0[i,j])) }
prod_2_k0[i] = (gamma(sum(alpha_init_0[i,]))/gamma(sum(count_init_0[i,] +
 alpha_init_0[i,])))
prod_k0 [i] = prod_1_k0[i]*prod_2_k0[i]}
for(i in 1:dias){
term_general_like[i, 1] = prod_k0[i]}
#######
# Caso K = 1
#######
# the initial states in this case is the first row of the case K=0
# assigning the values of count_init_0[1,i] to count_init_1[i]
count_init_1 <- matrix(0,1,2)
for(j in 1:2){count_init_1[j]= count_init_0[1,j]}
# assigning the same values of alpha_init_0[1,i] to alpha_init_1[i]
alpha_init_1 <- matrix(0,1,2)
for(j in 1:2){alpha_init_1[j]= alpha_init_0[1,j]}
#############
# counting the number of transitions 0 -> 0, 0 -> 1, 1 -> 0 and 1 -> 1
# count_0_k1[i,1] counts the number of transitions from zero to zero in row i,
```

```
# count_0_k1[i,2] counts the number of transitions from zero to one in row i,
#count_1_k1[i,1] counts the number of transitions from one to zero in row i
# count_1_k1[i,2] counts the number of transitions from ne to one in row i
##############
count_0_k1 <- matrix(0, dias-1, 2)
count_1_k1 <- matrix(0, dias-1, 2)
for(i in 1:(dias-1)){
for(j in 1:anos){
if((ozonio_sw[i, j] == 0) & (ozonio_sw[i+1,j] == 0))
{count_0_k1[i,1] = count_0_k1[i,1] + 1}
if((ozonio_sw[i, j] == 0) & (ozonio_sw[i+1,j] == 1))
{count_0_k1[i,2] = count_0_k1[i,2] + 1}
if((ozonio_sw[i, j] == 1) & (ozonio_sw[i+1,j] == 0))
{count_1_k1[i,1] = count_1_k1[i,1] + 1}
if((ozonio_sw[i, j] == 1) & (ozonio_sw[i+1,j] == 1))
{count_1_k1[i,2] = count_1_k1[i,2] + 1}}}
# assigning the values of the values of alpha
# alpha_0_k1 is associated with the transitions 0 -> 0 and 0 -> 1
# alpha_1_k1 is associated with the transitions 1 -> 0 and 1 -> 1
alpha_0_k1 <- matrix(0, dias-1, 2)
alpha_1_k1 <- matrix(0, dias-1, 2)
for(i in 1:(dias-1)){
for(j in 1:2){
if(count_0_k1[i,j] == 0){alpha_0_k1[i,j] = 3}
if(count_1_k1[i,j] == 0){alpha_1_k1[i,j] = 3} }
if((count_0_k1[i,1] == count_0_k1[i,2]) & (count_0_k1[i,1] != 0) &
(count_0_k1[i,1] < 5) & (alpha_0_k1[i, 1] == 0)){alpha_0_k1[i,1] = 5
if(alpha_0_k1[i,2] == 0){alpha_0_k1[i,2] = 5} }
if((count_1_k1[i,1] == count_1_k1[i,2]) & (count_1_k1[i,1] != 0) &
(count_1_k1[i,1] < 5) & (alpha_1_k1[i, 1] == 0)){alpha_1_k1[i,1] = 5
if(alpha_1_k1[i,2] == 0){alpha_1_k1[i,2] = 5} }
if((count_0_k1[i,1] == count_0_k1[i,2]) & (count_0_k1[i,1] != 0) &
(count_0_k1[i,1] >= 5) & (alpha_0_k1[i, 1] == 0)){alpha_0_k1[i,1] = 7
if(alpha_0_k1[i,2] == 0){alpha_0_k1[i,2] = 7} }
if((count_1_k1[i,1] == count_1_k1[i,2]) & (count_1_k1[i,1] != 0) &
(count_1_k1[i,1] >= 5) & (alpha_1_k1[i, 1] == 0)){
alpha_1_k1[i,1] = 7
if(alpha_1_k1[i,2] == 0){alpha_1_k1[i,2] = 7} }
if((count_0_k1[i,1] == min(count_0_k1[i,])) & (count_0_k1[i,1] != 0)){
if(alpha_0_k1[i,1] == 0){alpha_0_k1[i,1] = 4}
if(alpha_0_k1[i,2] == 0){alpha_0_k1[i,2] = 8} }
if((count_1_k1[i,1] == min(count_1_k1[i,])) & (count_1_k1[i,1] != 0)){
if(alpha_1_k1[i,1] == 0){alpha_1_k1[i,1] = 4}
if(alpha_1_k1[i,2] == 0){alpha_1_k1[i,2] = 8} }
if((count_0_k1[i,1] == min(count_0_k1[i,])) & (count_0_k1[i,1] == 0)){
if(alpha_0_k1[i,2] == 0){alpha_0_k1[i,2] = 8} }
if((count_1_k1[i,1] == min(count_1_k1[i,])) & (count_1_k1[i,1] == 0)){
if(alpha_1_k1[i,2] == 0){alpha_1_k1[i,2] = 8} }
if((count_0_k1[i,1] == max(count_0_k1[i,])) & (count_0_k1[i,2] != 0)){
if(alpha_0_k1[i,1] == 0){alpha_0_k1[i,1] = 8}
if(alpha_0_k1[i,2] == 0){alpha_0_k1[i,2] = 4} }
if((count_1_k1[i,1] == max(count_1_k1[i,])) & (count_1_k1[i,2] != 0)){
if(alpha_1_k1[i,1] == 0){alpha_1_k1[i,1] = 8}
if(alpha_1_k1[i,2] == 0){alpha_1_k1[i,2] = 4} }
if((count_0_k1[i,1] == max(count_0_k1[i,])) & (count_0_k1[i,2] == 0)){
if(alpha_0_k1[i,1] == 0){alpha_0_k1[i,1] = 8} }
if((count_1_k1[i,1] == max(count_1_k1[i,])) & (count_1_k1[i,2] == 0)){
if(alpha_1_k1[i,1] == 0){alpha_1_k1[i,1] = 8} }
}
# calculating the value of the likelihood L(Y | K = 1)
# term corresponding to the initial distribution
```

```
prod_1_k1 = 1
for(j in 1:2){
prod_1_k1 = prod_1_k1*((gamma(count_init_1[j]+alpha_init_1[j])/
gamma(alpha_init_1[j])))*((gamma(sum(alpha_init_1))/
gamma(sum(count_init_1+ alpha_init_1))))}
term_init_like[2] = prod_1_k1
# term corresponding to the rest of the days (rows)
prod_2_k1 <- matrix(0,(dias - 1),1)
prod_3_k1 <- matrix(0,(dias - 1),1)
prod_k1 <- matrix(0,(dias - 1),1)
for(i in 1:(dias - 1)){
for(j in 1:2){
prod_2_k1[i] = (gamma(count_0_k1[i,j] + alpha_0_k1[i,j])/gamma(alpha_0_k1[i,j]))*
(gamma(count_1_k1[i,j] + alpha_1_k1[i,j])/gamma(alpha_1_k1[i,j]))}
for(j in 1:2){
prod_3_k1[i] = (gamma(sum(alpha_0_k1[i,])+1)*gamma(sum(alpha_1_k1[i,])))/
(gamma(sum(count_0_k1[i,]+alpha_0_k1[i,])+sum(count_1_k1[i,]+alpha_1_k1[i,])+1))}
prod_k1[i] = prod_2_k1[i] * prod_3_k1[i]}
for(i in 1:(dias-1)){
term_general_like[i, 2] = prod_k1[i]}
####
# Case K >=2
#####
# Transforming vector of zeros and ones in an element of S_{2}^{(K)}
# counting the initial values of the chain (K >= 2) initialisation of the
# matrices to store the values of the likelihood for each order and day
##########
prod_init <- matrix(0, 1, M)
for(K in 2:M){
count_init_k <- matrix(0, 1, (2^K))
mbase_init <- matrix(0, 1, anos)
alpha_init_k <- matrix(0, 1, 2^K)
# transforming the k-dimensional initial vector into an integer number
for(j in 1:anos){
for(l in 0:(K-1)){
mbase_init[j] = mbase_init[j] + ozonio_sw[1+l,j]*2^l}
# counting the number of m in the initial state
for(n in 0:(2^K-1)){if(mbase_init[j] == n){
count_init_k[mbase_init[j]+1] = count_init_k[mbase_init[j]+1] + 1}}}
# assigning the respective values of alpha_init_k
for(n in 0:(2^K - 1)){
if(count_init_k[n+1] == 0){
alpha_init_k[n+1] = 3}
if((count_init_k[n+1] == min(count_init_k)) & (alpha_init_k[n+1] == 0)){
alpha_init_k[n+1] = 3}
if((count_init_k[n+1] == max(count_init_k)) & (alpha_init_k[n+1] == 0)){
alpha_init_k[n+1] = 8}
if((abs(count_init_k[n+1]-max(count_init_k)) > 5) & (alpha_init_k[n+1] == 0)){
alpha_init_k[n+1] = 4}
if((abs(count_init_k[n+1]-max(count_init_k)) <= 5) & (alpha_init_k[n+1] == 0))
{alpha_init_k[n+1] = 6}}
# calculation of the first term in the product in the likelihood L(Y | K)
prod_1_init = 1
prod_2_init = 1
prod_init_k = 0
for(n in 1:(2^K)){
prod_1_init = prod_1_init*(gamma(count_init_k[n]+alpha_init_k[n])
/gamma(alpha_init_k[n]))}
prod_2_init = 1
sumalphacount = sum(alpha_init_k+count_init_k)
sumalpha = sum(alpha_init_k)
```

```
limsup = sum(alpha_init_k+count_init_k) - sum(alpha_init_k)
for(k in 0:(limsup-1)){prod_2_init = prod_2_init*((sumalphacount-1)-k)^(-1)}
prod_init_k = prod_1_init*prod_2_init
term_init_like[K+1] = prod_init_k
# initialisation of the matrices of interest in the case of i not the initial state
mbase <- matrix(0, dias - K, anos) # that is overline{m}
count_Km <- matrix(0, 2^K, 2) # matrix counting the transitions from m
s <- matrix(0, (dias-K), 2^K)
# transforming the vectors of length K into a number in the base 2 for
# each day for all years
for(i in 1:(dias-K)){
for(j in 1:anos){
for(l in 0:(K-1)){
mbase[i,j] = mbase[i,j] + ozonio_sw[i+l,j]*2^l}
# storing the number of mbase in row i in the vector s[day, mbase]
for(n in 0:(2^K-1)){
if(mbase[i,j] == n){s[i,n+1] = s[i, n+1] + 1} }
} # closes the j loop
} # closes the i loop
# counting the number of transitions for each day (day is kept fixed while
# counting goes through years)
# count_Km[m,1] counts the number of transitions m -> 0 in the 22 years for fixed i
# count_Km[m,2] counts the number of transitions m -> 1 in the 22 years for fixed i
# n_m0[dias,m] counts the number of transitions m -> 0 in the 22 years for each i
# n_m1[dias,m] counts the number of transitions m -> 1 in the 22 years for each i
n_m0 <- matrix(0, (dias - K), 2^K) # matrix counting m -> 0
n_m1 <- matrix(0, (dias - K), 2^K) # matrix counting m -> 1
for(i in 1:(dias-K)){
for(j in 1:anos){
if(ozonio_sw[i+K,j] == 0){
for(n in 0:(2^K-1)){
if(mbase[i,j] == n){
count_Km[n+1,1] = count_Km[n+1,1]+1}}}
if(ozonio_sw[i+K,j] == 1){
for(n in 0:(2^K-1)){if(mbase[i,j] == n){
count_Km[n+1,2] = count_Km[n+1,2]+1}}}   }
for(m in 1:(2^K)){
n_m0[i,m] = count_Km[m,1]
n_m1[i,m] = count_Km[m,2]  }
count_Km <- matrix(0, 2^K, 2)
} # closes de i loop
# assignation of the values of the corresponding values of
# alpha the hyperparameter of the Dirichlet prior distribution
# alpha_m0 is associated to the transitions m -> 0 for each day and each m
# alpha_m1 is associated to the transitions m -> 1 for each day and each m
alpha_m0 <- matrix(0, (dias - K), 2^K)
alpha_m1 <- matrix(0, (dias - K), 2^K)
for(i in 1:(dias-K)){
for(m in 1:2^K){
if(n_m0[i,m] == 0){alpha_m0[i,m] = 3}
if(n_m1[i,m] == 0){alpha_m1[i,m] = 3}
} #closes the m loop
for(m in 1:2^K){
if((n_m0[i,m] == min(n_m0[i,m], n_m1[i,m])) & (alpha_m0[i,m] == 0)){
alpha_m0[i,m] = 4
if((abs(n_m0[i,m] - n_m1[i,m]) >= 5) & (alpha_m1[i,m] == 0)){alpha_m1[i,m] = 7}
if((abs(n_m0[i,m] - n_m1[i,m]) < 5) & (alpha_m1[i,m] == 0)){alpha_m1[i,m] = 5}}
if((n_m0[i,m] == min(n_m0[i,m], n_m1[i,m])) & (alpha_m0[i,m] != 0)){
if((abs(n_m0[i,m] - n_m1[i,m]) >= 5) & (alpha_m1[i,m] == 0)){alpha_m1[i,m] = 7}
if((abs(n_m0[i,m] - n_m1[i,m]) < 5) & (alpha_m1[i,m] == 0)){alpha_m1[i,m] = 5} }
if((n_m0[i,m] == max(n_m0[i,m], n_m1[i,m])) & (alpha_m0[i,m] == 0))
```

```
{ alpha_m0[i,m] = 8
if((abs(n_m0[i,m] - n_m1[i,m]) >= 5) & (alpha_m1[i,m] == 0)){alpha_m1[i,m] = 4}
if((abs(n_m0[i,m] - n_m1[i,m]) < 5) & (alpha_m1[i,m] == 0)){alpha_m1[i,m] = 7} }
if((n_m0[i,m] == min(n_m0[i,m], n_m1[i,m])) & (alpha_m0[i,m] != 0)){
if((abs(n_m0[i,m] - n_m1[i,m]) >= 5) & (alpha_m1[i,m] == 0)){alpha_m1[i,m] = 4}
if((abs(n_m0[i,m] - n_m1[i,m]) < 5) & (alpha_m1[i,m] == 0)){alpha_m1[i,m] = 7} }
} #closes the second m loop
} #closes the i loop
# calculation of the likelihood L(Y | K)
prod_1_km <- matrix(0, (dias - K), 1)
prod_2_km <- matrix(0, (dias - K), 1)
prod_km <- matrix(0, (dias-K), 1)
for(i in 1:(dias-K)){
for(n in 1:2^K){
prod_1_km[i] = (gamma(n_m0[i,m]+alpha_m0[i,m])*gamma(n_m1[i,m]+alpha_m1[i,m]))/
(gamma(alpha_m0[i,m])*gamma(alpha_m1[i,m]))
prod_2_km[i] = gamma(alpha_m0[i,m] + alpha_m1[i,m])/gamma(alpha_m0[i,m] +
n_m0[i,m] + n_m1[i,m])
} #closed the m loop
prod_km[i] = prod_1_km[i]*prod_2_km[i]
} # closes the i loop
for(i in 1:(dias-K)){
term_general_like[i,K+1] = prod_km[i]}
} #close the K loop
write.csv(term_general_like, file = "results-file-1.csv")
write.csv(term_init_like, file = "results-file-2.txt")
#
# TRANSITION PROBABILITIES - CASE OF REGION CE, THRESHOLD 0.17
#
K = 8 # may change for other regions and thresholds
# calculation of the normalising constant in the case of the initial distribution
somainit = 0
for(m in 1:2^K){
somainit = somainit + (alpha_init_k[m] + count_init_k[m] - 1)}
# calculation of the initial distribution
prob_init <- matrix(0, 2^K, 1)
for(m in 1: 2^K){
prob_init[m] = (alpha_init_k[m] + count_init_k[m] - 1)/somainit}
write.csv(prob_init, file = "prob_init_chain.txt")
# limiting the number of days
dd = 2 # because I need p_{mj}(1) and p_{mj}(2)
# alpha_m1[i,m] where i is day and m is the value of the vector
# the m1 indicates that m is followed by 1
p_m0 <- matrix(0, dd, 2^K)
p_m1 <- matrix(0, dd, 2^K)
# calculation of the transition probabilities m -> 0 and m -> 1
for(m in 1:2^K){
for(i in 1:dd){
p_m0[i,m] = (alpha_m0[i,m] + n_m0[i,m] - 1)/((alpha_m1[i,m] + n_m1[i,m] - 1)
+(alpha_m0[i,m] + n_m0[i,m] - 1))
p_m1[i,m] = (alpha_m1[i,m] + n_m1[i,m] - 1)/((alpha_m1[i,m] + n_m1[i,m] - 1)
+(alpha_m0[i,m] + n_m0[i,m] - 1)) }}
trans_mat <- matrix(0, 2^K, dd)
for(m in 1:2^K){
for(i in 1:dd){
trans_mat[m,i] = p_m0[i,m]}}
write.csv(trans_mat, file = "trans_mat.txt")
```

Author details

Eliane R. Rodrigues[1*], Mario H. Tarumoto[2] and Guadalupe Tzintzun[3]

1 Instituto de Matemáticas, Universidad Nacional Autónoma de México, Mexico

2 Faculdade de Ciênicas e Tecnologia, Universidade Estadual Paulista Júlio de Mesquita Filho, Brazil

3 Instituto Nacional de Ecología y Cambio Climático, Secretaría de Medio Ambiente y Recursos Naturales, Mexico

*Address all correspondence to: eliane@math.unam.mx

References

[1] Bell ML, McDermontt A, Zeger SL, Samet JM, Dominici F. Ozone and short-term mortality in 95 US urban communities, 1987-2000. Journal of the American Medical Society 2004; 292: 2372-2378.

[2] Bell ML, Peng R, Dominici F. The exposure-response curve for ozone and risk of mortality and the adequacy of current ozone regulations. Environmental Health Perspectives 2005; 114: 532-536.

[3] Cifuentes L, Borja-Arbuto VH, Gouveia N, Thurston G, Davis DL. Assessing the health benefits of urban air pollution reduction associated with climate change mitigation (2000-2020): Santiago, São Paulo, Mexico City and New York City. Environmental Health Perspectives 2001; 109: 419-425.

[4] Dockery DW, Schwartz J, Spengler JD. Air pollution and daily mortality: association with particulates and acid aerosols. Environmental Research 1992; 59: 362-373.

[5] Galizia A, Kinney PL. Long-term residence in areas of high ozone: association with respiratory health in a nationwide sample of nonsmoking adults. Environmental Health 1999; 99: 675-679.

[6] Gauderman WJ, Avol E, Gililand F, Vora H, Thomas D, Berhane K, McConnel R, Kuenzli N, Lurmman F, Rappaport E, Margolis H, Bates D, Peter J. The effects of air pollution on lung development from 10 to 18 years of age. The New England Journal of Medicine 2004; 351: 1057-1067.

[7] Gouveia N, Fletcher T. Time series analysis of air pollution and mortality: effects by cause, age and socio-economics status. Journal of Epidemiology and Community Health 2000; 54: 750-755.

[8] Loomis D, Borja-Arbuto VH, Bangdiwala SI, Shy CM. Ozone exposure and daily mortality in Mexico City: a time series analysis. Health Effects Institute Research Report 1996; 75: 1-46.

[9] Martins LC, de Oliveira Latorre MRD, Saldiva PHN, Braga ALF. Air pollution and emergency rooms visit due to chronic lower respiratory diseases in the elderly: an ecological time series study in São Paulo, Brazil. J. Occupational and Environmental Medicine 2002; 44: 622-627.

[10] WHO. Air Quality Guidelines-2005, Particulate Matter, Ozone, Ditrogen dioxide and Sulfur Dioxide. European Union: World Health Organization Regional Office for Europe; 2006.

[11] Horowitz J. Extreme values from a nonstationary stochastic process: an application to air quality analysis. Technometrics 1980; 22: 469-482.

[12] Smith RL. Extreme value analysis of environmental time series: an application to trend detection in ground-level ozone. Statistical Sciences 1989; 4: 367-393.

[13] Raftery AE. Are ozone exceedance rate decreasing?. Comment of the paper *Extreme value analysis of environmental time series: an application to trend detection in ground-level ozone by R. L. Smith*. Statistical Sciences 1989; 4: 378–381.

[14] Flaum JB, Rao ST, Zurbenko IG. Moderating Influence of Meteorological Conditions on Ambient Ozone Concentrations. Journal of the Air and Waste Management Association 1996; 46: 33-46.

[15] Achcar JA, Rodrigues ER, Tzintzun G. Using stochastic volatility models to analyse weekly ozone averages in Mexico City. Environmental and Ecological Statistics 2011; 18: 271-290.

[16] Javits JS. Statistical interdependencies in the ozone national ambient air quality standard. Journal of Air Pollution Control Association 1980; 30: 58-59.

[17] Achcar JA, Fernández-Bremauntz AA, Rodrigues ER, Tzintzun G. Estimating the number of ozone peaks in Mexico City using a non-homogeneous Poisson model. Environmetrics 2008; 19: 469-485.

[18] Achcar JA, Rodrigues ER, Tzintzun G. Using non-homogeneous Poisson models with multiple change-points to estimate the number of ozone exceedances in Mexico City. Environmetrics 2011; 22: 1-12.

[19] Villaseñor-Alva JA, González-Estrada E. On modelling cluster maxima with applications to ozone data from Mexico City. Environmetrics 2010; 21: 528-540.

[20] Barrios JM, Rodrigues ER. A queueing model to study occurrence and duration of ozone exceedances in Mexico City. Journal of Applied Statistics 2014. dx.doi.org/101080/02664763.2014.939613

[21] Rajagopalan B, Upmanu L, Tarboton DG. Non-homogeneous Markov model for daily precipitation. Journal of Hydrology Engineering 1996; 1: 33-40.

[22] Hughes JP, Guttorp P, Charles SP. A non-homogeneous hidden Markov model for precipitation occurrence. Applied Statistics, Part 1 1999; 48: 16-30.

[23] Drton M, Marzban C, Guttorp P, Schafer JT. A Markov chain model for tornadic activity. American Meteorological Society 2003; 131: 2941-2953.

[24] Larsen LC, Bradley RA, Honcoop GL. A new method of characterizing the variability of air quality-related indicators. Air and Waste Management Association. International Specialty Conference, Tropospheric Ozone and the Environment. Pittsburgh, USA: California Air and Waste Management Series; 1990.

[25] Álvarez LJ, Fernández-Bremauntz AA, Rodrigues ER, Tzintzun G. Maximum a posteriori estimation of the daily ozone peaks in Mexico City. Journal of Agricultural, Biological, and Environmental Statistics 2005; 10: 276-290.

[26] Álvarez LJ, Rodrigues ER. A trans-dimensional MCMC algorithm to estimate the order of a Markov chain: an application to ozone peaks in Mexico City. International Journal of Pure and Applied Mathematics 2008; 48: 315-331.

[27] Cox DR, Lewis PA. Stochastic analysis of series of events. UK: Methuen; 1966.

[28] Rice JA. Mathematical statistics with data analysis. New York, USA: Wadsworth and Brook; 1988.

[29] Carlin BP, Louis TA. Bayes and Empirical Bayes Methods for Data Analysis. Second Edition. USA: Chapman and Hall/CRC; 2000.

[30] Green PJ. Reversible jump Markov chain Monte Carlo computation and Bayesian model determination. Biometrika 1995; 82: 711-732.

[31] Carlin BP, Chib S. Bayesian model choice via Markov chain Monte Carlo methods. Journal of the Royal Statistical Society Series B 1995; 57: 473-484.

[32] Boys RJ, Henderson DA. On determining the order of a Markov dependence of an observed process governed by a hidden Markov chain. Special Issue of Scientific Programming 2002; 10: 241-251.

[33] Fleming TR, Harrington DP. Estimation for discrete time non-homogeneous Markov chains. Stochastic Processes and their Applications 1978; 7: 131-139.

[34] Aalem OO, Johansen S. An empirical transition matrix for non-homogeneous Markov chains based on censored observation. Scandinavian Journal of Statistics 1978; 5: 141-150.

[35] Robert CP, Casella G. Monte Carlo statistical methods. New York, USA: Springer; 1999.

[36] Fan T-H, Tsai C-A. A Bayesian method in determining the order of a finite Markov chain. Communications in Statistics – Theory and Methods 1999; 28: 1711-1730.

[37] Evans M, Swartz T. Approximating Integrals via Monte Carlo and Deterministic Methods. Oxford Statistical Series 20. Oxford, UK: Oxford University Press; 2000.

[38] NOM. Modificación a la Norma Oficial Mexicana NOM-020-SSA1-1993. Diario Oficial de la Federación. 30 Octubre 2002. Mexico: 2002. (In Spanish.)

5

Emission Control Technology

Thanh-Dong Pham, Byeong-Kyu Lee,
Chi-Hyeon Lee and Minh-Viet Nguyen

Additional information is available at the end of the chapter

1. Introduction

The chapter presents various promising methods to control air pollution emissions. Although several previously published books have examined this field [1-6], most of them presented only the methods to control air pollutants generated from stationary sources. These days, however, mobile sources are also important sources contributing to air pollution. Among the six air criteria pollutants listed by the United States Environmental Protection Agency (EPA), including CO, particulate matter (PM), SO_2, NO_2, O_3, and Pb, mobile sources can contribute to 81% of all CO emissions, 37% of PM, 4% of SO_2, 45% of NO_2, 47% of volatile organic compounds (VOCs are precursors of O_3, their estimate is a surrogate of the O_3 concentration), and 72% of Pb [6]. Therefore, promising methods to control the emission of air pollutants generated from mobile sources should be included in the chapter to provide the readers with innovative ideas about the emission control of air pollution. Because of the variety of mobile sources and their mobility from one location to another, the methods applied to control the emission of air pollutants generated from stationary sources may not be useful for controlling air pollutants generated from mobile sources. Therefore, in this chapter, upstream control strategies, which try to control air pollutants from upstream emission processes, will be presented as promising methods to control the emission of air pollutants generated from mobile sources. Policies and regulations applying to control air pollutants emitted from transportation activities, agricultural activities, and construction fields will be presented as the main strategies to air pollution from upstream processes.

On the other hand, the chapter also presents downstream technologies to control air pollutants emitted from stationary sources. Based on the characteristics of target pollutants, the downstream control technologies will be classified into two categories: particulate matter and gaseous pollutant control technologies. The technologies of cyclone, wet scrubber, electrostatic precipitators, and fabric filter will be introduced in the chapter as methods to control particulate matter pollutants. To control gaseous pollutants, the chapter will present methods such as adsorption, absorption, condensation, incineration, applications of biological system, and photocatalyst. Each technology will be presented in detail from definition to principle and applications. The advantages and disadvantages of each technology will be described and compared in the chapter.

2. Upstream control strategies

2.1. Control transportation activities

Strategies to control air pollutants emitted from transportation activities include regulations to control precursor pollutants in raw materials; the application of catalytic converters to reduce NO_x, CO, and hydrocarbon emissions; the control of lubricant consumption; the reduction of motorized transportation demand; and the improvements to road quality and traffic flow.

2.1.1. Raw materials

Air pollutants can be controlled based on regulations that control precursor pollutants in raw materials such as sulfur and lead. Sulfur oxides (SO_x) in aerosol are formed during the combustion process of sulfur in fossil fuels such as gasoline and diesel. Therefore, the use of low-sulfur-content fuels can be considered as a control strategy to reduce SO_2 emissions. Many regulations have been issued to reduce the sulfur content in all transportation fuels. For example, Figure 1 presents a selection of a few of the gasoline and diesel sulfur specifications in major countries and the regulatory timetable associated with the introduction of these specifications [7].

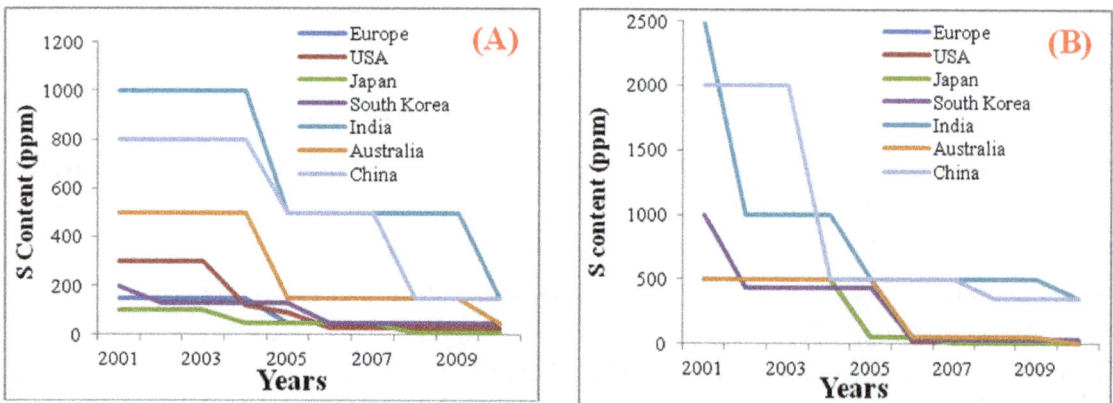

Figure 1. Regulations and regulatory timetable for sulfur content in gasoline (A) and diesel (B) in several countries.

In the 1920s, refiners started adding lead compounds to gasoline in order to increase octane levels and improve engine performance by reducing engine "knock" and allowing higher engine compression. However, the burning of leaded fuel introduced massive quantities of atmospheric lead leading to many adverse effects on human health. Therefore, regulations to eliminate lead from leaded fuel began to be issued worldwide from the 1970s. For example, the US EPA scheduled performance standards requiring refineries to decrease the average lead content of all fuels beginning in 1975, but these were postponed until 1979 through a series of regulatory adjustments. By the early 1980s, the lead content in fuels had declined by about 80% due to both the regulations and the fleet turnover [8]. In August 1984, the US EPA proposed a further reduction of lead to 0.1 grams per liter gallon (gplg) by January 1, 1986. However, several refineries were not able to achieve this time scale, so the US EPA postponed the deadline until January 1, 1988. Lead was banned as a fuel additive in the United States beginning in 1996. Figure 2 shows the decline over time in the lead content of leaded fuel in the United States [9].

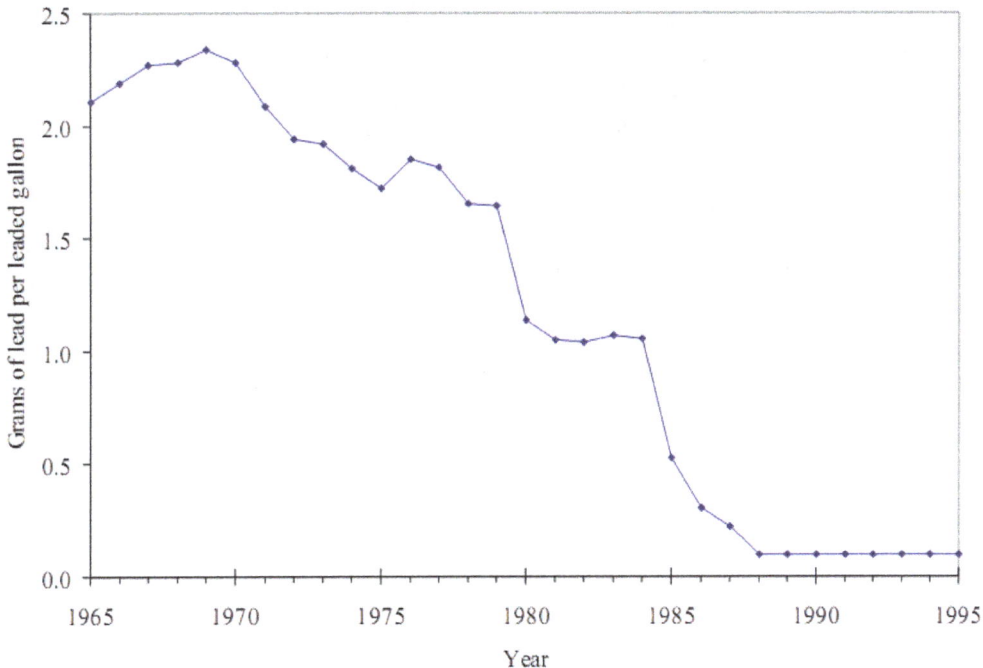

Figure 2. Temporal lead content in leaded fuel in the United States.

Regulations concerning oxygenated gasoline and reformulated gasoline were also issued in the United States in 1990 to reduce the vehicular emissions of CO and VOCs. The oxygenated gasoline regulation required a higher oxygen content in gasoline in order to ensure more complete burning of the gasoline and thereby reduce harmful tailpipe emissions from motor vehicles. In this respect, oxygen dilutes or displaces the precursor pollutant components in gasoline such as aromatics (e.g., benzene) and sulfur and thus decreases their content in gasoline. The reformulated gasoline regulation is specially blended gasoline. Therefore, the gasoline can be burned more cleanly and can be prevented from evaporating as quickly as conventional gasoline, thereby reducing the emission of smog-forming and toxic pollutants. To reduce the emission of benzene (VOCs) into an aerosol, the US EPA required all refiners to meet an annual average gasoline benzene content standard of 0.62% by volume (vol %) in all their gasoline, both reformulated and conventional, nationwide, from 2011.

In addition, the use of new-generation fuels such as biofuels and natural gas can be a valuable strategy to reduce air pollutant emission. Feedstocks for biofuels, mostly plants, are much more environmental-friendly and evenly distributed around the world than the feedstocks for traditional fuels such as oil and gas. Two of the most commonly used biofuels are ethanol and biodiesel. Biodiesel fuels are oxygenated organic compounds of methyl or ethyl esters derived from a variety of renewable sources such as vegetable oil, animal fat, and cooking oil. Therefore, the use of biofuels for transportation activities can significantly reduce the atmospheric emissions of CO, hydrocarbon, and lead.

2.1.2. Catalytic converters

A catalytic converter is a vehicle emission control device that converts toxic pollutants in exhaust gas to less-toxic pollutants by catalyzing a redox reaction (oxidation or reduction). A catalytic converter comprised usually of the following three main parts is used in internal combustion engines: substrate, washcoat, and catalytic materials. The substrate material is usually a ceramic monolith with a honeycomb structure. A washcoat, usually aluminum oxide, titanium dioxide, silicon dioxide, or a mixture of silica and alumina, is used as a carrier for the catalytic materials. The catalytic material is often a mix of precious metals such as platinum, palladium, and rhodium. The catalytic converter can reduce oxides of nitrogen (NO_x) into nitrogen gas (N_2), combine or oxidize carbon monoxide (CO) with/or unburned hydrocarbons (HC) to produce carbon dioxide (CO_2) and water (H_2O). A schematic diagram illustrating the role of the catalytic converter in reducing air pollutants is shown in Figure 3.

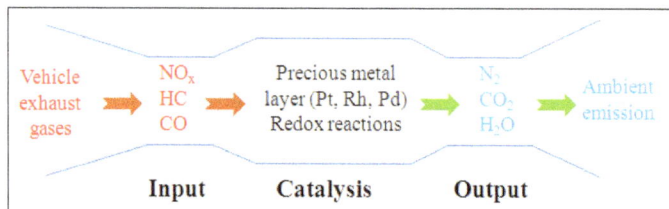

Figure 3. A schematic diagram of a catalytic converter.

The relevant catalysts applied to reduce the pollutants emitted from specific engines are listed in Table 1 [10].

To reduce NO_x, a controlled amount of the reactive chemical reductant such as anhydrous ammonia, aqueous ammonia, and urea is added to a stream of fuel or exhaust gas and adsorbed onto the catalyst. Due to the role of the catalysts, NO_x can react with the reductant to produce nitrogen gas and water according to the following reactions:

Engines	Pollutants	Catalysts
Gasoline engines	Nitrogen oxides (NO_x), Hydrocarbons	$Pt/Pd/Rh/Ce_xZr_{1-x}O_2/(La, Ba)-Al_2O_3$ on ceramic
Diesel engines (light vehicles)	(HC_s), and Carbon monoxide (CO)	and metallic monoliths
Diesel engines (heavy vehicles,	NO_x, HC_s, CO	$Pt/Pd/Rh/BaO/Al_2O_3$ on ceramic and metallic
truck, and bus)	NO_x, HC_s, CO	monoliths
Diesel engines	Particulate matter	V_2O_x/TiO_2 on ceramic monolith
		Cerium and iron oxides
		Pt/Al_2O_3
		Cu, V- and K-based catalysts

Table 1. The relevant catalysts for specific engines

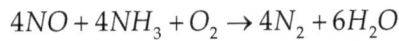

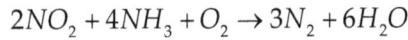

$$4NO + 4NH_3 + O_2 \rightarrow 4N_2 + 6H_2O$$
$$2NO_2 + 4NH_3 + O_2 \rightarrow 3N_2 + 6H_2O$$
$$NO + NO_2 + 2NH_3 \rightarrow 2N_2 + 3H_2O$$

The reaction typically takes places at an optimal temperature range between 630 and 720 K, but can operate from 500 to 720 K with longer residence times. To operate an effective process, the engine requires an external urea tank and dosing system. The specific NO_x/ammonia and ammonia/catalyst ratios can be designed to optimize a specific application. At optimal conditions, the application of a catalyst in the downstream of engine can reduce NO_x emissions from vehicles by 70–90%.

The reactions to oxidize carbon monoxide and unburned hydrocarbon are described by the following reactions:

$$2CO + O_2 \rightarrow 2CO_2$$
$$C_xH_{2x+2} + [(3x+1)/2]O_2 \rightarrow xCO_2 + (x+1)H_2O$$

2.1.3. Lubricant consumption

Lubricants are composed of a base fluid and additives. The base fluid is the major part of the lubricant formulation and is mainly made from petroleum-based oils. The additives are used to obtain desirable properties. Lubricants are very important substances for reducing the wear and tear of machine parts. Lubricants also reduce friction, which in turn reduces heat loss. The worldwide consumption of lubricants is more than 41 million tones [11]. They have a soluble organic fraction of 60%, which contributed between 20% and 90% of the total particulates in air that were generated from engine lubricant consumption. Therefore, the particulate emission rate can be significantly reduced by controlling engine lubricant consumption. The strategies to control engine lubricant consumption include changing the piston-ring design and manipulating the operation conditions of the engine such as intake air pressure. The use of biolubricants is also a valuable strategy to reduce the adverse effects of traditional lubricants to the atmospheric emissions of particulate matter. Biolubricants are described in many ways such as eco-friendly lubricants, green lubricants, biodegradable lubricants, recyclable, nontoxic, and reusable.

2.1.4. Reduce motorized transportation demand

Strategies applied to encourage the use of nonmotorized transport, discourage nonessential trips, shorten trip lengths, and restrain the use of private cars can reduce the overall demand for motorized transport and thus minimize the emission of air pollutants from transportation activities. The strategies could be instigated based on the following regulations:

- Provide safe and comfortable conditions for walking and other forms of nonmotorized transport
- Improve public transportation quality and efficiency
- Increase fuel taxes
- Increase parking charges
- Increase road pricing
- Compact design of retail and entertainment centers with workers and public transport
- Limit use of private vehicles both by pricing and by administrative regulation

2.1.5. Road quality and traffic flow

Road quality also directly affects the air pollutants emitted from transportation activities. For example, the operation of vehicles on unpaved roads can introduce a significant amount of atmospheric particulate matter. Therefore, road qualities need to be improved to reduce air pollutant emission. The strategies to improve road quality include:

- Try to pave unpaved roads
- Sweep roads frequently (can reduce concentrations of PM up to 20%)
- Flush roads with water in the dry season
- Investigate new types of asphalt and concrete, which are cheap and environmental-friendly
- Cover operating trucks

Strategies to improve traffic flow and thereby minimize unnecessary braking and reduce congestion can result in high efficiency of vehicle operation and reduce undesired pollutant emissions. These strategies can be obtained based on the following methods:

- Control traffic signals
- Design road systems by use of ring roads and bypasses
- Increase infrastructure capacity
- Reduce congestion by congestion charging
- Reduce vehicle speeds because fast moving vehicles stir up dust (a reduction in speed from 40 miles per hour (m/h) to 20 (m/h) reduces dust emissions by 65%)

2.2. Control agricultural activities

Air pollutants emitted from the agricultural sector are mainly methane (CH_4), nitrous oxide (N_2O), and ammonia (NH_3). Agriculture is also a major source of PM, both primary and secondary in origin [12]. Agricultural pollutants are mainly generated from livestock production and the application of fertilizers and pesticides. Therefore, strategies to control air pollutants emitted from agriculture activities are strongly linked to the activities including strategies to control livestock feeding, animal housing systems, manure storage systems, application of manure for crops, and application of fertilizers and pesticides.

2.2.1. Livestock feeding

Because the quantity of nitro compounds, such as ammonia and nitrous oxide excreted from animal feces and urine, is linearly dependent on the intake of nitrogen in food (protein), the strategies to reduce the oversupply of protein in animal feedstock can reduce nitrogen excretions and thus decrease the emissions of nitrogen-containing compounds [13]. Such strategies involve adapting the amount of proteins in the food to the needs of the animals. For instance, young animals and high-productive animals require more protein than older and less-productive animals. On average, this measure leads to a NH_3 emission reduction of 10% for a 1% reduction in the mean protein content in the diet, but efficiencies depend strongly on the animal categories. It has no implications on animal health as long as the requirements for all amino acids are ensured. It is most applicable to housed animals while the practical applicability of feeding strategies to grazing animals is limited.

2.2.2. Animal housing systems

The available strategies to reduce NH_3 emissions from animal housing systems have been well known for decades and apply one or more of the following principles [13]:

Principles	NH_3 emission reduction
· Decrease the surface area fouled by manure	15–25% in pig housing
· Rapid removal of urine and rapid separation of feces and urine	25–46%
· Reduce pH of the manure	up to 20%
· Reduce temperature of the manure	up to 60% in pig and cattle housing systems
· Dry the manure (e.g., poultry litter)	45–75%
· Scrubbing ammonia from exhaust air	up to 70%
· Decrease housing time by increasing grazing time	70–95%
	10–50%, but some emission swapping

2.2.3. Manure storage

The strategies to eliminate air pollutants emitted from manure storage can be based on the following principles:

- Decreasing the surface area where emissions can take place, i.e., by covering the storage, encouraging crusting, and increasing the storage depth

- Reducing the pH and temperature of the manure

- Minimizing disturbances such as aeration

2.2.4. Manure used for crops

The application of manure for crops can emit a significant amount of atmospheric pollutants. The strategies or application techniques to control the emissions can be based on the following principles:

- Decrease the exposed surface area of slurries applied to surface soil through band application, injection, and incorporation

- Decrease the time that emissions can take place, i.e., bury the slurry or solid manures through injection or incorporation into the soil

- Decrease the source strength of the emitting surface, i.e., through lowering the pH and NH_4 concentration of the manure (through dilution)

2.2.5. Fertilizer application

The strategies to reduce emissions of pollutants from the application of fertilizers are based on one or more principles including:

- Decrease emission sites by decreasing the surface area via band application, injection, and incorporation

- Decrease the emission periods of pollutant via rapid incorporation of fertilizers into the soil or via irrigation

- Decrease the emitting source surface strength via urea inhibitors and blending

- Ban use of pollutant precursors such as ammonium (bi) carbonate

For example, the techniques for the application of urea and ammonium-based fertilizers can reduce levels of ammonia emission as follows [13]:

Fertilizer type	Application techniques	Emission reduction %
Urea	Injection	>80
	Urea inhibitors	>30
	Incorporation following surface application	>50
	Surface spreading with irrigation	>40
	Ban	~100
	Injection	>80
Ammonium carbonate	Incorporation following surface application	>50
Ammonium-based fertilizers	Surface spreading with irrigation	>40

2.3. Control construction fields

The strategies to control air pollutants emitted from construction fields can be classified into four categories: control of site planning, construction traffic, demolition works, and site activities [14].

2.3.1. Site planning

Regulations applied to control air pollutants emitted from site planning include:

- Erect effective barriers around dusty activities or entire site boundary
- Do not burn any material in entire site planning
- All site personnel to be fully trained
- Trained and responsible manager on site during working times to maintain logbook and carry out site inspections
- Hard surface all major haul routes through the site (e.g., use recycled rubber blocks, concrete blocks or tarmac)
- Use nearby rail or waterways for transportation to/from site

2.3.2. Construction traffic

Regulations applied to control air pollutants emitted from construction traffic activities include:

- No vehicles or plant will be left idling unnecessarily
- Wash or clean all vehicles effectively before leaving the site
- All loads entering and leaving site should to be covered
- On-road vehicles comply to emission standards
- Use ultra-low sulfur tax-exempt diesel for all nonroad mobile machineries
- Should minimize construction traffic activity around site
- Cover the haul routes with hard surface combining with frequently cleaning the surface and give an appropriate control of speed limit around site

2.3.3. Demolition works

Regulations applied to control air pollutants emitted from demolition works include:

- Use water as dust suppressant during demolition works
- Use enclosed chutes and covered skips
- Wrap building(s) before demolition works
- Bag and remove any biological debris or damp down before demolition
- Avoid explosive blasting where possible and consider using appropriate manual or mechanical alternatives

2.3.4. Site activities

Regulations applied to control air pollutants emitted from site activities include:

- Minimize dust-generating activities. For example, when a worker cuts concrete slabs or bricks with a power tool without extraction or suppression, a second worker can pour water from a plastic bottle over the material leading to reduce the great amount of generated dust.

- Use water as a dust suppressant where applicable

- Cover, seed, or fence stockpiles to prevent wind whipping

- Re-vegetate earthworks and exposed areas

- If applicable, ensure the concrete crusher or concrete batcher has a permit to operate

- Minimize drop heights to control the fall of materials

2.4. Miscellaneous

Coal, the most abundant solid fuel and widely used for power plant and other industrial activities, is the largest source of air pollutant emissions. Coal combustion produces a significant amount of air pollutants such as SO_x, heavy metals, and PM. For example, sulfur in coal occurs both as inorganic minerals (mainly pyrite and marcassite) and organic compounds incorporated in the combustible part of coal. The sulfur content can be converted into SOx during the coal combustion. Therefore, reducing the sulfur content in coal before the combustion processes is a great strategy to reduce SO_x emissions from the upstream coal combustion process. Inorganic sulfur in coal can be removed by coal washing and the organic sulfur by using chemical hydrogenation and gasification processes.

Mining activities can produce significant air pollutants such as heavy metals (in PM form), SO_x and NO_x. Strategies to reduce air pollutants emitted from mining activities from the upstream process include enclosure or cover mine, mining area, and transfer areas; water spraying mining area; and stabilizing unpaved traffic areas.

Indoor activities can also be a significant source of air pollution. Strategies to reduce air pollutants emitted from indoor activities include improvement of cooking devices, use of alternative fuels for cooking and reducing the need for fire. Strategies to improve cooking devices include stabilization of stove materials and improvement of stove chimneys, in particular, biomass stoves. Uses of alternative fuels for cooking including charcoal, biogas, liquid petroleum gas, and electricity can significantly reduce air pollutant emissions. For example, the transition from wood to charcoal for cooking can reduce PM10 emissions by more than 80% (although the wider environmental impacts of charcoal production must be considered). The need for fire can be reduced based on the use of solar heating or electric devices.

The change of building materials from high-polluting materials such as paint, linoleum, and gypsum to low-polluting materials such as PVC and polyolefin can also control air pollutants from upstream emissions.

3. Downstream control

3.1. Particulate matter control

3.1.1. Cyclone

The cyclone is a well-known device used primarily for the collection of medium-sized and coarse particles. The cyclone works by forcing a gaseous suspension downward. The particles move outward by centrifugal force and collide with the outer wall and then slide downward to the bottom of the cyclone. At the bottom of the cyclone, the gas reverses its downward spiral and moves upward in a smaller, inner spiral. The cleaned air exits from the top of the cyclone and the particles are expelled from the bottom of the cyclone through a pipe sealed by a spring-loaded flapper valve or a rotary valve. The cyclone collector is shown schematically in Figure 4.

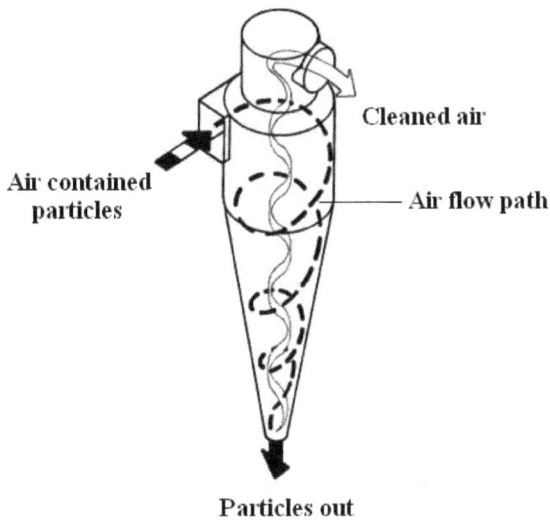

Figure 4. Schematic diagram of a cyclone.

Cyclones have a wide range of industrial applications in gaseous cleaning and product recovery. They are relatively inexpensive, easy to set up and maintain, and can work at high temperature and pressure. They can be used as a precollector for removing larger particles before next treatment. When well designed, the cyclone can collect particles larger than 10 μm with an efficiency of more than 90%. For smaller particles, however, the well-designed cyclone would have a considerable pressure drop with relatively lower collection efficiency [15]. In addition, the cyclone method cannot be used for removing sticky particles with high moisture content.

3.1.2. Wet scrubber

A wet scrubber system can be used to control fumes, mists, acid gasses, heavy metals, trace organics, and suspended dusts. An individual wet scrubber can usually be used to control a targeted pollutant. Therefore, a well-designed wet scrubber system often contains two or more single scrubbers leading to a multistage wet scrubber, which affords higher total removal efficiencies than that of a single-stage scrubber [1]. A schematic diagram of a wet scrubber is shown in Figure 5.

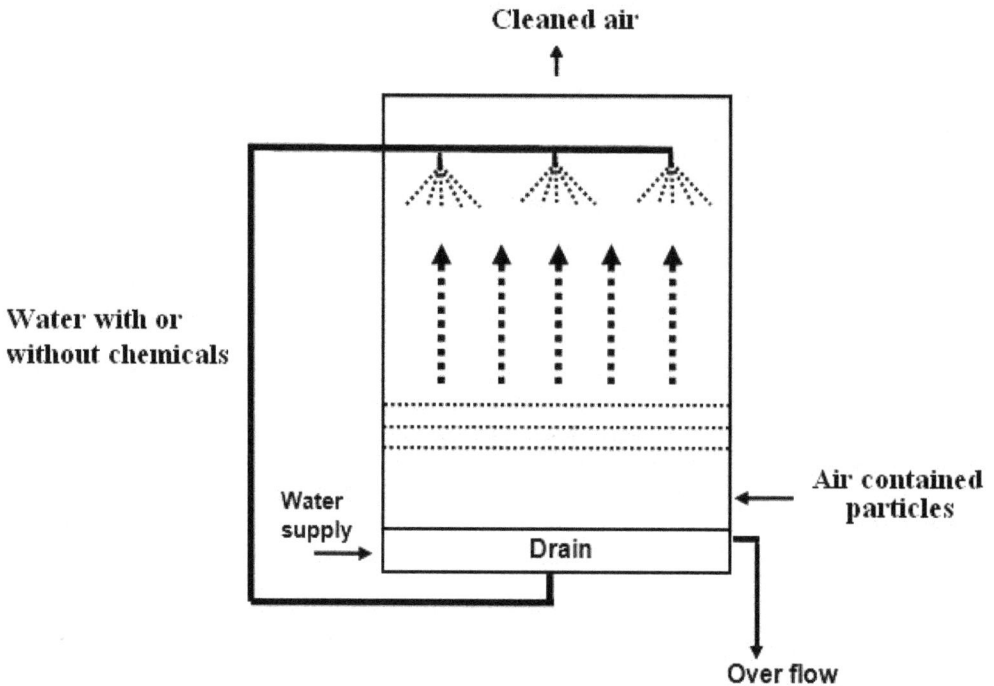

Figure 5. Schematic diagram of a wet scrubber chamber.

The wet scrubber system works based on direct interaction between the adsorbent liquid and the particles. The adsorbent liquid is usually water; however, several chemicals are also added to water to increase the adsorption ability of the liquid phase with the particles. Based on the interaction between the particles and the liquid phase, the particles can diffuse out of the gas phase and be absorbed in the liquid phase and the particle-loaded air can be cleaned. The absorption of particles into the liquid phase can be both physical and chemical absorption, depending on the particle and liquid phase and gaseous properties. The particle removal efficiency depends on:

- The solubility of the pollutant in the chosen scrubbing liquor

- Pollutant concentration in the gas phase being treated

- Flow rate of the gas and liquid phases

- Gas–liquid phase contact surfaces

- Stripping efficiency of the liquor and recycling of the solvent

- The ability to increase gas–liquid phase contact surfaces will result in higher absorption efficiency in a wet scrubber system. The reduced temperature and increased liquid-to-air ratio improve the absorption efficiency of a wet scrubber system.

- The pH of the scrubbing liquid is an important factor directly affecting the purification efficiency. The pH of the liquid may need to be low for ammonia scrubbing, while a neutral or high pH liquid is required for acid gas scrubbing. When scrubbing trace organics, a wet scrubber with alkaline pH is often used.

3.1.3. Electrostatic precipitators

Electrostatic precipitation (ESP), which is one of the most popular and efficient particle control systems in the United States, is defined as a particle control method that uses electrical forces to move the particles out of the flowing gas stream and onto collector plates [16]. The ESP processes include:

1. The ionization of particles, which can be dry dusts or liquid droplets, in contaminated air (particle charging)

2. The charged particles are deposited on an oppositely charged plate

3. The removal of the deposited particles from the plates

The particles are charged when the particles in the air stream pass through a corona, a region of gaseous ions flow. In the corona, the ions bombard the surface of the particles leading to charging particles. When these charged particles pass through the surface of the collecting electrodes, oppositely charged plates, they are trapped on the collected electrodes by the electrostatic field. The charged particles are accelerated toward the collecting electrodes by Coulomb forces, but inertial and viscous forces can resist the motion. When the plates (electrodes) collected a certain particle amount, the collected particles must be removed from the plates to prevent their re-entrainment into the gas stream. The plates could be knocked to let the collected layer of particles to slide down into a hopper from which they are evacuated [16]. The plates could also be continuously washed with water to remove the collected particles. A schematic diagram of an electrostatic precipitator is shown in Figure 6.

The principal difference between the ESP and other scrubbing methods are that in the ESP, the separation forces are electrical and are applied directly to the particles or droplets themselves while in others the separation forces are usually applied indirectly through the contaminated air system [17]. Therefore, the ESP could remove small particles or liquid droplets at a high efficiency with low energy consumption or low cost and small pressure drop through the gas cleaning system [17]. ESPs are built in either single-stage or two-stage versions. In the single-stage precipitator, the ionization and collection of particles or liquid droplets are achieved in a single stage and the corona discharge and precipitating field extend over the full length of the device. In the two-stage precipitator, the ionization of particles or liquid droplets is carried out in the first stage confined to the region around the corona discharge wires, followed by the particle collection in the second stage which provides an electrostatic field whereby the previously charged particles migrate onto the surface of the collecting electrodes [2].

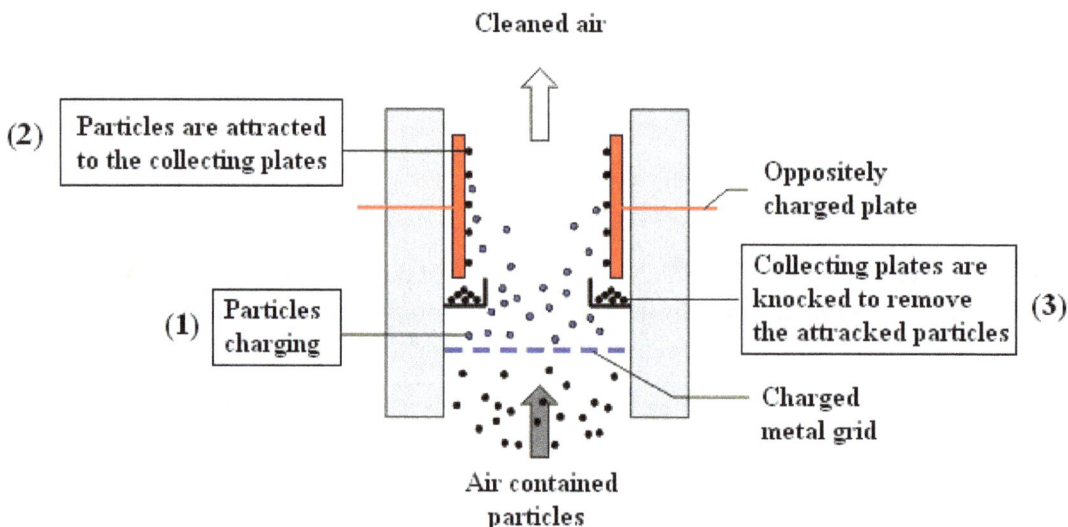

Cleaned air

(2) | Particles are attracted to the collecting plates

Oppositely charged plate

Collecting plates are knocked to remove the attracked particles | (3)

(1) | Particles charging

Charged metal grid

Air contained particles

Figure 6. Schematic diagram of an electrostatic precipitator.

3.1.4. Fabric filtration

Fabric filtration is a well known and accepted physical technique in which a gas stream containing mainly solids passes through a porous fabric medium which retains the solids. This process may operate in a batch or semicontinuous mode for removing the retained solid particles from the filter medium. Filtration systems may also be designed to operate in a continuous manner.

In air fabric filtration, the contaminated gas flows into and passes through a number of filter bags placed in parallel, leaving the solid particles retained by the fabric filter. The fabric filter can be classified into two basic groups, depending on the fabric properties: felt and woven. Felt media are normally used in high-energy cleaning systems, while woven media are used in low-energy systems. Felt fabrics are tighter in construction and they can be considered to be more of a true filter medium and should be kept as clean as possible to perform satisfactorily as a filter. The woven fabric is merely a site upon which the true filtering occurs as the dust layer builds up, through which the actual filtering take place.

Particles are collected on the fabric surface through four mechanisms including:

- Inertial collection – the fibers, which are placed perpendicular to the gas flow direction, could collect the particles in the stream without changing gas flow direction.

- Interception – particles are trapped in the filter matrix.

- Brownian movement – submicron particles are collected on the surface of the filter.

- Electrostatic forces – the particles on the gas stream were captured because of electrostatic interaction between the particles and the filter.

The particles were captured on the filter leading to formation of a dust cake on the filter. The formation of the dust cake could increase the resistance to gas flow. Therefore, the filter containing the dust cake must be frequently cleaned.

Fabric filters are extremely efficient solid removal devices and operate at nearly 100% efficiency. The efficiency depends on several factors including:

- Particle properties

  o Size: Particles between 0.1 and 1.0 μm in diameter may be more difficult to capture.

  o Seepage characteristics: Small, spherical solid particles tend to escape.

  o Inlet dust concentration: The deposit is likely to seal over sooner at high concentrations.

- Fabric properties

  o Surface depth: Shallow surfaces form a sealant dust cake sooner than napped surfaces do.

  o Weave thickness: Fabrics with high permeability, when clean, show lower efficiencies.

  o Electrostatics: Particles, fabrics, and gas can all be influenced electrostatically and proper combination.

- Dust cake properties

  o Residual weight: The heavier the residual loading, the sooner the filter is apt to seal over.

  o Residual particle size: The smaller the base particles, the smaller (and fewer) are the particles likely to escape.

- Air properties

  o Humidity: With some dusts and fabrics, 60% relative humidity is much more effective than 20% relative humidity. Increased humidity or moisture level can be a frequent cause of clogging pores of the filter medium and increasing filter pressure drop.

- Operational variables

  o Velocity: Increased velocity usually lowers the efficiency, but this can be reversed depending on the collection mechanisms, for example, impaction and infusion.

  o Pressure: Probably not a factor, except that an increase in pressure after the dust cake has been formed can fracture the filter medium and greatly reduce efficiency until the cake reseals.

  o Cleaning: Without frequent or periodical cleaning, the air filtration system cannot be operated.

The advantages and disadvantages of methods to control particulate matter including cyclone, wet scrubber, ESP, and fabric filtration are summarized in Table 2.

Advantages	Disadvantages
Cyclone	
· Low capital cost	· Relatively low efficiencies for collection particles which
· Simple and insignificant maintenance problems	are smaller than 10 mm
· Ability to operate at high temperature	· Could not use for sticky materials
· Require small spaces	
Wet scrubber	
· No secondary production	· Treatment issue concerning with water disposal/effluent
· Require small space	· Corrosion problems
· Operation to collect both gases and sticky particles	· High pressure drop problems
· Operation at high-temperature as well as high-humidity	· Solid buildup problems at the wet–dry interface
gas streams	· Relatively high maintenance costs
· Low capital cost	
· Operation with flammable and explosive dust with little	
risk	
· High effective to collect fine particles	
Electrostatic precipitation (ESP)	
· Very high collection efficiencies of coarse as well as fine	· High capital cost
particulates with low energy consumption	· High sensitivity to fluctuations in gas stream
· Collection dry dust	· Problems with particles, with extremely high or low
· Low pressure drop	resistivity
· Relatively low operation and maintenance costs	· Require large space for installation
· Operation capability at high temperatures as well as high	· Produce ozone as by-product
pressure or under vacuum	· Require highly trained maintenance personnel
· High collect capacity	
Fabric filtration	
· Very high collection efficiencies of coarse as well as fine	· Difficult to operate at high temperature
particulates	· Need for fabric treatment after removal process
· Relative insensitivity to gas stream fluctuations and large	· Require high maintenance costs
changes in inlet dust loadings	· Explosion problems
· Recirculation of filter outlet air	· Shortened fabric life at elevated temperatures and in the
· No corrosion issues	presence of acid or alkaline particulate or gas
· Simple maintenance, flammable dust collection	· Respiratory protection requirement for fabric
· High collection efficiency of submicron smoke and	replacement
gaseous contaminants	· Medium pressure-drop requirements
· Many application types	
· Simple operation	

Source: Bounicore and Davis 1992 [18]

Table 2. Advantages and disadvantages of particulate control methods

3.2. Gaseous pollutants control

3.2.1. Adsorption

Adsorption is the phenomenon via which molecules of a fluid adhere to the surface of a solid material (adsorbent). Gas adsorption is used for industrial applications such as odor control, recovery of volatile solvents such as benzene, toluene, and chloroflurocarbon, and drying of process gas streams. During this process, the molecules or particles (adsorbate) in airstream gases and liquids can be selectively removed or captured despite being at low concentrations. There are two distinct adsorption mechanisms: physisorption and chemisorption. Physisorption or physical adsorption, also called van der Waals adsorption, involves a weak bonding of gas molecules with the adsorbent. The bond energy is similar to the attraction forces between molecules in the stream. The adsorption process is exothermic and the heat of adsorption is slightly higher than the heat of the vaporization of the adsorbate. The forces holding the adsorbate to the adsorbent are easily overcome by either the application of heat or the reduction of pressure, which are methods that can be used to regenerate the adsorbent. Chemisorption or chemical adsorption involves an actual chemical bonding by reaction of the adsorbate with the adsorbent, leading to new chemical bonds such as covalent bonding generated at the adsorbent surface.

When a stream comes into contact with an adsorbent, one or several components of the stream are adsorbed by the adsorbent. At all adsorbent interfaces, adsorption can occur, but often at a low level unless the adsorbent is highly porous and possesses fine capillaries. For an effective solid adsorbent, it should have a large surface-to-volume ratio, and a preferential affinity for the individual component of concern. The adsorption occurs by a series of steps. In the first step, the adsorbate diffuses from the stream to the external surface of the adsorbent. In the second step, the adsorbate molecule migrates from the relatively small area of the external surface to the pores within each adsorbent. The bulk of the adsorption occurs in these pores because of the majority of available surface area. In the final step, the adsorbate adheres to the surface in the pores of the adsorbent [19].

Most industrial adsorbents could be divided into three classes including:

- Oxygen-containing compounds such as silica gel and zeolite

- Carbon-based compounds such as graphite or activated carbon materials

- Polymer-based compounds, which include functional groups in a porous polymer matrix

Silica gel, which is usually prepared by the reaction between sodium silicate and acetic acid, is a chemically inert, nontoxic, polar, and dimensionally stable amorphous form of SiO_2. It is used for the drying of processed air and the adsorption of polar hydrocarbons from natural gas.

Zeolites are natural or synthetic crystalline aluminosilicates, which have a repeating pore network and release water at high temperature. Zeolites are applied in the drying of processed air, CO_2 removal from natural gas, CO removal from reforming gas, air separation, catalytic cracking, and catalytic synthesis and reforming.

Activated carbon is a highly porous, amorphous solid consisting of microcrystallines with a graphite lattice, usually prepared in small pellets or a powder. It is nonpolar and cheap. Activated carbon is used for the adsorption of organic substances and nonpolar adsorbate. Activated carbon is also usually used for waste gas (and waste water) treatment. It is the most widely used adsorbent because its chemical and physical properties such as surface groups, pore size distribution, and surface area can be tuned as required. Its usefulness also derives from its large microspore (and sometimes mesoporous) volume and the resulting high surface area.

3.2.2. Absorption

Absorption is a physical or chemical process in which atoms, molecules, or ions enter some bulk phase – gas, liquid, or solid material. As compared to the adsorption process, in which the molecules are adhered on the surface of the adsorbent, the absorption process takes place when the volume takes up molecules.

Gas absorption is the removal of one or more pollutants from a contaminated gas stream when the gas stream passes through a gas–liquid interface and ultimate dispersion in the liquid. Absorption is a process that may be chemical (reactive) or physical (nonreactive). Physical absorption is formed based on the interaction of two phases of matter including a liquid absorbs a gas or a solid absorbs a liquid. When a liquid solvent absorbs a part or all of a gas mixture, the gas mass could move into the liquid volume. The mass transfer could take place at the interface between the gas and the liquid. The mass transfer rate depends on both the liquid and the gas properties. The solubility of gases, the pressure and the temperature are the main factors affecting to this type of absorption. In addition, the absorption rate also depends on the surface area of the interface and its duration in time. When a solid absorbs a part or all of a liquid mixture, the liquid mass could move into the solid volume. The mass transfer could take place at the interface between the liquid and the solid. The mass transfer rate depends on both the solid and the liquid properties. Chemical absorption or reactive absorption is a chemical reaction between the absorbed and the absorbing substances. Sometimes, it is combined with physical absorption. This type of absorption depends upon the stoichiometry of the reaction and the concentration of its reactants.

Gas absorption is usually carried out in packed towers. The contaminated gas stream enters the bottom of the column and passes upward through a wetted packed bed. The absorbing liquid enters from the top of the column and is distributed over the column packing. The column packing may have one or more commercially available geometric shapes designed to maximize the gas–liquid contact and minimize the gas–phase pressure drop [20]. The requirements for the packing column include high wetted area per unit volume, minimal weight, sufficient chemical resistance, low liquid holdup, and low pressure drop.

3.2.3. Condensation

Condensation is a separation process to convert one or more volatile components of a vapor mixture to a liquid through saturation process. Any volatile components can be converted to liquids by sufficiently lowering their temperature and increasing their pressure. The most common process is reducing the temperature of the vapor because increasing the vapor pressure is expensive. The condensation process is primarily used to remove VOCs from gas streams prior to other control methods, but sometimes it can be used alone to reduce emissions from high-VOCs concentration gas streams [2, 21].

The simple and relatively inexpensive condenser uses water or air to cool and condense the vapor stream to the liquid. Since these devices are not required to reach or capable of reaching low temperature, high removal efficiencies of most vapor pollutants cannot be obtained unless the vapor will condense at high temperature. That is why condensers are typically used as a pretreatment device. They can be used together with adsorption, absorption, and incinerators to reduce the gas volume to be treated by other expensive methods.

A typical condenser device includes condenser, refrigeration system, storage tanks, and pumps. The condensation process includes:

- The contaminated gas stream is compressed as it passes through the blower.

- The existing hot gas stream flows to an after-cooler, which is constructed of copper tubes with external aluminum fins. Air is passed over the fins to maximize the cooling effect. Some condensation occurs at this step.

- The gas stream continues to cool in an air-to-air heat exchanger.

- The condenser cools the gas to below the condensing temperature in an air-to-refrigerant heat exchanger.

- Finally, the cold gas passes to a centrifugal separator where the liquid is removed to the collecting vessel. The gas stream typically requires further treatment before being emitted to the atmosphere.

A condensing system usually contains either a contact condenser or a surface condenser. Contact condensing systems cool the contaminated gas stream by spraying ambient or chilled liquid directly into the gas stream. A packed column is usually used to maximize the surface area and contact time. The direct mixing of the coolant and contaminant necessitates separation or extraction before coolant reuse. This separation process may lead to a disposal problem or secondary emissions. Contact condensers usually remove more contaminated air as a result of greater condensate dilution. In the surface condensing systems, the coolant does not mix with the gas stream, but flows on one side of a tube or plate in the surface condensing systems. The condensing vapor contacts the other side, forms a film on the cooled surface, and drains into a collection vessel for storage, reuse, or disposal. Surface condensers require less water and generate 10–20 times less condensation than contact condensers do.

The advantages of the condensation method include lower installation cost, little required auxiliary equipment, and less maintenance requirement. However, the remaining disadvantages of the method include problems of water disposal, low efficiencies, and the need for further treatment.

3.2.4. Incineration

Incineration or thermal oxidation is a broadly used method to control air pollutants such as VOCs, using oxidation at high temperature. Incineration is considered as an ultimate disposal technique in which VOCs are converted to carbon dioxide, water, and other inorganic gases. The two popular incineration methods are thermal incineration and catalytic incineration.

In thermal incineration, organic compounds in the contaminated gas are burned or oxidized at a high temperature with air in the presence of oxygen [22]. The thermal oxidizer involves specifying a temperature of operation along with a desired residence time and then optimum sizing the device to achieve the desired residence time and temperature with proper flow velocity. Selection of the proper piece of equipment depends on the mode of operation, oxygen content, and concentration of the organic gases. They are very important when trying to minimize the overall cost of the incineration and reduce the volume of the gas stream to be treated as much as possible. Depending on the types of heat recovery, incinerators can be classified into two categories: recuperative and regenerative. The recuperative incinerator uses a shell and tube heat exchanger to transfer the heat generated by the incinerator to the preheat of the feed stream. The recuperative incinerator can recover about 70% of the waste heat from the exhaust gases [21]. The regenerative incinerator includes a flame-based combustion chamber that connects two or three fixed beds containing ceramic or other inert packing. The input gas enters over these beds where it is preheated before passing into the combustion chamber and being burned. Then, the hot flue gases pass through the packed beds where the heat generated during incineration is recovered and stored. The packed beds keep the heat during one cycle and release it as the beds preheat the input organic gases in the second cycle. This regenerative incinerator method can recover up to 95% of the energy from the flue gas [23].

In catalytic incineration, the organic compounds in the contaminated gas are converted into carbon dioxide and water by using a catalyst that facilitates incineration at low temperature. Thus, the requirement incineration temperature can be decreased by hundreds of degrees. Therefore, the application of catalyst incineration can save a large amount of energy to heat up the gas stream containing pollutants for combustion. The contaminated gases are heated by a small auxiliary burner, and then the gases passed through the catalyst bed. The space requirement for operation of catalytic incineration is much smaller than that of thermal incineration. Thanks to the catalytic activity, the degree of oxidation of the pollutants is greatly increased compared with that in the incineration system without any catalyst. The catalyst activity refers to the degree of the chemical reaction rate. The catalyst can also be selective with higher activity for some compounds. Such activity and selectivity enable a lower operating temperature while still achieving the desired destruction efficiencies. In air pollution control, the catalyst is usually a noble metal (Pd, Pt, Cr, Mn, Cu, Co, and Ni) deposited on an alumina support in a configuration to minimize the pressure drop, which is often critical for incinerator designs [23, 24].

3.2.5. Biological system

The biological system for controlling air pollutants such as VOCs and odor uses microbes or microorganisms, immobilized on a biologically active solid support, to treat the gas pollutants.

The principle of the method is that the gaseous pollutants are used by the microbes as a food or energy source and thus destroyed and converted into innocuous metabolic end products such as carbon dioxide and water. The processes via which microbes destroy or convert pollutants contain:

- First, the pollutant gas must be absorbed into the liquid film in which the microbes are growing.

- The pollutant is absorbed into the cells of the microbes and metabolized.

- Finally, the end products (mainly CO_2) must be expelled from the cell and diffuse outward through the liquid film.

The process requires careful attention to design and operation in order to ensure firstly good contact between the contaminated gases and the microbes contained on the solid support; and secondly, that the microbe population is sustained and maintained in a healthy state. The key concerns in the design and operation of a biological system to control air pollutants contain [25]:

- Identify the type of the contaminant and its concentration in the air stream

- Find the suitable microbial population and combining with a compatible medium

- Maintain sufficient moisture

- Design the bed to ensure suitable residence time for a given airflow rate

- Control the bed conditions including pH, nutrient levels, and temperature

The biological technology is most suitable for high volumetric flow rate air streams containing low pollutant concentrations. The two most common biological systems are biofilter and bioscrubber. Biofilter is a biological system which uses an organic or synthetic media to host and nourish the microorganism without the requirement for an aqueous flush system. Bioscrubber is a biological system which uses an inorganic or synthetic media to provide a structural base for physically hosting the microorganisms requiring a continuous water flush or an intermittent containing carbon nutrient that supports the microorganism. The use of a biological system to control air pollutant offers several advantages including effective removal of compounds, little or no by-product pollutants, uncomplicated installations, and low costs. However, the method retains several disadvantages including a reduced suitability to high concentration streams, large area requirement for installation, need for careful attention to moisture control, and the possibility of becoming clogged by particulate matter or biomass growth.

3.2.6. Application of photocatalyst

An alternative technology, which offers a number of advantages over the above-mentioned technologies, for controlling organic air pollutants is to use photocatalysts. The use of a photocatalyst supports the operation of a low- or room-temperature photocatalytic oxidation process that can degrade a broad range of organic contaminants into innocuous final products such as CO_2 and H_2O without significant energy input. When the photocatalyst, for example, TiO_2, is irradiated with ultraviolet (UV) radiation that has energy higher than the band gap energy of the TiO_2, an electron could be excited from the valence band to the conduction band of TiO_2, leaving holes behind on the TiO_2 surface. The leaved holes could react with surrounding H_2O to produce hydroxyl radicals ($^*OH$), while the excited electrons could react with O_2 to produce superoxide radical anions $\left(o_2^-\right)$. These oxy radical species can participate in the oxidation reaction to destroy many organic contaminants ($C_xH_yO_z$) completely [26]. The photocatalytic process mainly follows the following reactions:

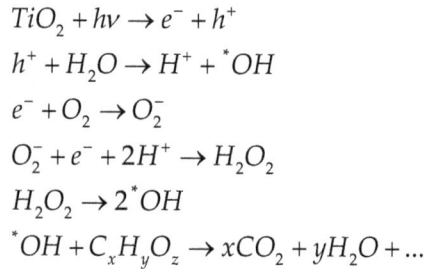

$$TiO_2 + h\nu \rightarrow e^- + h^+$$
$$h^+ + H_2O \rightarrow H^+ + {^*OH}$$
$$e^- + O_2 \rightarrow O_2^-$$
$$O_2^- + e^- + 2H^+ \rightarrow H_2O_2$$
$$H_2O_2 \rightarrow 2{^*OH}$$
$${^*OH} + C_xH_yO_z \rightarrow xCO_2 + yH_2O + \ldots$$

The three common reactor types designed to use a photocatalyst for air purification purposes are the honeycomb monolith, fluidized-bed, and annular reactors [27]. A honeycomb monolith reactor contains a certain number of channels, each of which typically has an internal dimension of the order of 1 mm. The cross-sectional shapes of the channels are square or circular. The photocatalyst is coated onto the walls of channels in a very thin wash coat. Fluidized-bed reactors are designed to treat a high gas feed rates directly passing through the catalyst bed. Based on reactor design, the solid photocatalyst could directly contact with the UV irradiation as well as gaseous reactants. The fluidized-bed reactors generally consisted of two concentric cylinders, which form an annular region with a certain gap. The photocatalyst is deposited onto the interior wall of the outer cylinder. The light source is usually located at the center. The thickness of the deposited photocatalyst film is sufficiently thin ensuring that all of the photocatalyst could be illuminated by UV irradiation [28].

The applications of the photocatalyst for photocatalytic oxidation processes to reduce air pollutants have been considered as alternatives to conventional air pollution control technologies. However, they have yet to overcome the problems of low energy efficiency and poor cost competitiveness. Therefore, numerous methods for modifying photocatalysts have been developed and investigated to accelerate the photo-conversion, enable the absorption of visible light, or alter the reaction mechanism to control the products and intermediates [29]. In this regard, metals or nonmetals were used as doping agents to implant or coprecipitate on the surface or in the lattice of TiO_2. Electron donors or hole scavengers have been added to such photocatalytic systems. In addition, another semiconductor was integrated with TiO_2 to establish a suitable two-semiconductor system [29]. The modifications not only change the mechanism and kinetics of the photocatalytic processes under UV irradiation but also enhance the photocatalytic activities of the photocatalyst, thereby enabling the photocatalytic oxidation processes to proceed even under visible light [30].

4. Strategies to control climate change

Global climate change, also called global warming or the greenhouse effect, may be the most significant problem ever faced by humankind. Global climate change is caused by adding certain gases including carbon dioxide (CO_2), methane (CH_4), nitrous oxide (N_2O), and many chlorofluorocarbons (CFCs). The added gases absorbed infrared radiation leading to excess thermal energy within the earth's biosphere. The largest driver of global warming is carbon dioxide (CO_2) emissions from fossil fuel combustion, cement production, and land use changes such as deforestation. Therefore, the strategies to control CO_2 emission are also the main strategies to control global climate change.

4.1. Upstream control

Upstream strategies to control CO_2 emission include energy conservation, alternative and renewable fuels, and oxy-fuel combustion.

4.1.1. Energy conservation and efficiency use

Energy conservation refers to reducing energy consumption through using less of an energy service. The strategies concerned to energy conservation include energy taxes, building design, transportation, and consumer products.

• Energy taxes

Some countries employ energy or carbon taxes to motivate energy users to reduce their consumption.

• Building design

Energy conservation in building could be improved by using of an energy audit, which is an inspection and analysis of energy use and flows in the building, process, or system to reduce the amount of energy input into the system without negatively affecting the output(s). In addition, a passive solar building design in which windows, walls, and floors are made to collect, store, and distribute solar energy in the form of light or heat in the winter and reject solar heat in the summer. The design, leading to decreased use of mechanical and electrical devices, is also a solution for energy conservation.

• Transportation

The zoning reform and designs for walking and bicycling could allow greater urban density leading to reduce energy consumption concerning to transportation. The application of telecommuting is also a sufficient opportunity to conserve energy. For example, with people who work in service jobs, they could work at home instead of commuting to work each day.

• Consumer products

Because the consumers usually lack the information concerning to saving by energy-efficient products, we must inform the consumer in understanding the problems. For example, many consumers choose cheap incandescent bulbs, failing to take into account their higher energy costs and lower lifespan. However, as compared to modern compact fluorescent and LED bulbs, these products have a higher upfront cost, with their long lifespan and low energy.

4.1.2. Alternative and renewable fuels

Use of alternative energy sources could prevent CO_2 emission from fossil fuel. The alternative energy sources include wind, solar, hydropower, biomass, and geothermal energy.

• Wind energy

Airflows can be used to run wind turbines to produce electric energy. Globally, the long-term technical potential of wind energy is believed to be five times total current global energy production, or 40 times current electricity demand.

• Solar energy

Solar energy, radiant light and heat from the sun, is harnessed using a range of ever-evolving technologies such as solar heating, photo-voltaic, concentrated solar power, solar architecture, and artificial photosynthesis.

• Hydropower

Hydropower is the power derived from the energy of falling water and running water, which may be harnessed for useful purposes. Hydropower could be also captured from ocean surface waves and tidal power.

• Biomass

Biomass is the biological material derived from living or recently living organisms. Biomass could be used as energy source by either used directly via combustion to produce heat or indirectly after converting it to various forms of biofuels.

• Geothermal energy

Geothermal energy could be generated from thermal energy, which is stored in the earth. Because of the difference between temperature of the core of the planet and its surface, it drives a continuous conduction of thermal energy in the form of heat from the core to the surface.

4.1.3. Oxy-fuel combustion

When burning coal or other fossil fuel using ambient air, the air contains a huge amount of nitrogen as well as the oxygen needed for combustion (4:1). Raising the temperature of the nitrogen to the combustion temperature requires a great deal of heat. Therefore, reducing nitrogen content in the air input could be a good strategy to reduce fuel consumption, leading to reduced CO_2 emission. In the strategy, oxy-fuel combustion, a process of burning a fuel using pure oxygen instead of air as the primary oxidant, is applied. There are several researches being done in firing fossil-fueled power plants with an oxygen-enriched gas mixture instead of air. Almost all of the nitrogen is removed from input air, yielding a stream that is approximately 95% oxygen.

4.2. Downstream control

Technologies to downstream control CO_2 or remove CO_2 from atmospheric include biological capture, wet scrubbing, CO_2 absorption, and mineral carbonation.

4.2.1. Biological capture

Biological capture of CO_2 is a process in which photosynthetic organisms are used to absorb the CO_2 gas from air and convert it into a solid carbonaceous compound. The strategies to conduct for biological capture of CO_2 include:

• Reforestation

Reforestation means that tree could be replanted on marginal crop and pasturelands leading to incorporate atmospheric carbon (CO_2) into biomass. For a successful process, the incorporated carbon could not return to the atmosphere from burning or rotting when the trees die. Finally, the trees grow in perpetuity or the wood from them must itself be sequestered in the forms of biochar or bio-energy with carbon storage or landfill.

• Agriculture

Agricultural activity to capture CO_2 is also called as "capture" energy of the sun. Under solar light, the artificial plants could capture and convert the CO_2 in atmosphere into biomass, which can be storage or used as food or also used as raw material to make biofuels to replace the use of fossil fuel. Land-based plants such as corn and soybeans can be grown as energy crops, in particular to make biodiesel. Because of the limitations of land-based plants, there has been much interest over the years in developing systems that utilize microalgae for engineered biological CO_2 capture systems. Microalgae can fix CO_2 up to ten times faster than trees, and utilize sunlight much more efficiently than the land-based energy crops.

4.2.2. Wet scrubbing of CO_2

A carbon dioxide scrubber, which uses various amines such as monoethanolamine as absorbents, could absorb CO_2 to capture them from the atmosphere. The design and principle of wet scrubber have been presented in Section 3.2.2. Amines could be used as absorbents to absorb CO_2 based on following reactions:

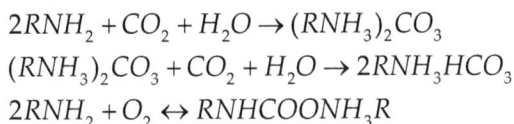

$$2RNH_2 + CO_2 + H_2O \rightarrow (RNH_3)_2CO_3$$
$$(RNH_3)_2CO_3 + CO_2 + H_2O \rightarrow 2RNH_3HCO_3$$
$$2RNH_2 + O_2 \leftrightarrow RNHCOONH_3R$$

4.2.3. Mineral carbonation

Many chemical processes, known as carbon sequestration by mineral carbonation or mineral sequestration, could capture and store CO_2 from the atmosphere in stable carbonate mineral forms. In the process, CO_2 could react with abundantly available metal oxides such as MgO or CaO to form stable carbonates. These reactions are mostly exothermic and occur naturally.

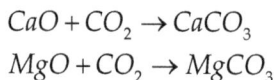

$$CaO + CO_2 \rightarrow CaCO_3$$
$$MgO + CO_2 \rightarrow MgCO_3$$

CO_2 could also react with calcium and magnesium silicates including forsterite and serpentinite by the following the reactions:

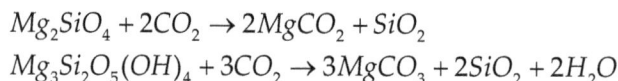

$$Mg_2SiO_4 + 2CO_2 \rightarrow 2MgCO_2 + SiO_2$$
$$Mg_3Si_2O_5(OH)_4 + 3CO_2 \rightarrow 3MgCO_3 + 2SiO_2 + 2H_2O$$

These reactions are slightly more favorable at low temperatures. This process occurs naturally over geologic time frames and is responsible for much of the earth's surface limestone.

5. Conclusion

In the chapter, we described the current and emerging technologies and strategies that are being used or proposed to control air pollutants. Control technologies and strategies could be classified into upstream and downstream controls. Upstream technologies and strategies are usually being used to control air pollution emitted from mobile sources such as transportation, agriculture, and construction activities. Downstream technologies are usually applied to control particulate matter and gaseous pollutants emitted from stationary sources. The upstream control exhibited more advantages concerning to cost and efficiency than the downstream control.

The air pollutants including SO_x and Pb emitted from transportation activities could be significantly reduced by controlling the precursor pollutants in raw materials. The NO_x, CO, and hydrocarbon emissions from transportation activity could be reduced by the application of catalytic converters and the control of lubricant consumption. The strategies to reduce motorized transportation demand and improve the road quality and traffic flow decrease the energy consumption demand, leading to decrease in air pollution emission. The strategies to control livestock feeding, animal housing systems, manure storage systems, application of manure for crops and application of fertilizers and pesticides significantly reduced the emission of methane (CH_4), nitrous oxide (N_2O), and ammonia (NH_3). The strategies to control construction activities including control of site planning, construction traffic, demolition works, and site activities significantly reduced the emission of particulate matter.

To control particulate matter pollutants, the technologies including cyclone, wet scrubber, electrostatic precipitators, and fabric filter have been efficiency applied. Among them, electrostatic precipitator is the most advantageous method and widely applied in the United

States. The traditional methods such as adsorption, absorption, condensation, incineration to control air pollutants emitted from stationary sources exhibited many certain disadvantages including high cost and efficiency. Therefore, they should to be modified, enhanced, or replaced by modern technologies such as biological and photocatalytic technologies.

Acknowledgements

This work was supported by a grant from the National Research Foundation of Korea (NRF), funded by the Ministry of Science, ICT, and Future Planning (2013R1A2A2A03013138).

Author details

Thanh-Dong Pham, Byeong-Kyu Lee*, Chi-Hyeon Lee and Minh-Viet Nguyen

*Address all correspondence to: bklee@ulsan.ac.kr

Department of Civil and Environmental Engineering, University of Ulsan, Daehakro , Namgu, Ulsan, Republic of Korea

References

[1] Cooper CD, Alley FC. Airpollution Control: A Design Approach. Waveland Press Inc, USA, 2011.

[2] Davis WT. Air Pollution Engineering Manual. John Wiley & Son, Inc, USA, 2000.

[3] Heinsohn RJ, Kabel RL. Sources and Control of Air Pollution. Prentice-Hall Inc, UK, 2000.

[4] Wang LK, Pereira NC, Hung Y-T. Air Pollution Control Engineering. Humana Press Inc, USA, 2010.

[5] Vallero D. Fundamentals of Air Pollution. 4 ed., Elsevier Inc, USA, 2008.

[6] Arcadio PS, Gregoria AS. Environmental Engineering: A Design Approach. Prentice-Hall, Inc, USA, 1996.

[7] Eric Y. Alternative sulfur management solutions to help refiners meet clean fuel and environmental challenges. In: 10th Biennial Sulfur Market Symposium. Bejjing, China, 2006.

[8] Nichols AL. Lead in gasoline. In: Economic Analyses at EPA: Assessing Regulatory Impact, Resources for the Future. 1997;1:49–86.

[9] Newell RG, Rogers K. The U.S. Experience with the Phasedown of Lead in Gasoline, in Washington, USA, 2003.

[10] Bradley MJ. Best available technology for air pollution control: Analysis guidance and case studies for North America in USA, 2005.

[11] Amit KJ, Amit S. Biotechnology: a way to control environmental pollution by alternative lubricants. Res Biotechnol 2013;4:38–42.

[12] Erisman JW, Bleeker A, Hensen A, Vermeulen A. Agricultural air quality in Europe and the future perspectives. Atmospheric Environment 2008;42:3209–17.

[13] Susana LA, Guerreiro C, Viana M, Reche C, Querol X. Contribution of agriculture to air quality problems in cities and in rural areas in Europe. In: The European Topic Centre on Air Pollution and Climate Change Mitigatio, 2013;10:1–26.

[14] Livingstone K. The Control of Dust and Emissions from Construction and Demolition. In: London, 2006.

[15] Leith D, Jones DL. Handbook of Power Science and Technology. 2 ed., Chapman & Hall, New York, USA, 1997.

[16] Treybal RE. Mass Transfer Operations, McGraw-Hill, New York, 1980.

[17] White HJ. Role of electrostatic precipitators in particulate control: a retrospective and prospective view. J Air Pollut Control Assoc 1975;25:102–7.

[18] Bounicore AJ, Davis WT. Air Pollution Engineering Manual. New York: Van Nostrand Reinhold, 1992.

[19] Glasstone S. Physical Chemistry. Princeton, USA, 1946.

[20] Calvert S, Englund HM. Handbook of Air Pollution Technology. Wiley-Interscience, New York, USA, 1984.

[21] Wang LK, Pereira NC, Hung YT. Air Pollution Control Engineering. Humna Press Inc, USA, 2010.

[22] Clausen CA, Copper CD, Hewett M, Martinez A. Enhancement of organic vapor incineraion using ozone. J Hazard Mater 1992;31: 75–87.

[23] Taylor PH, Delligner B, Lee CC. Development of a thermal stability based ranking of hazardous organic compound incinerability. Environ Sci Technol 1990;24:316–28.

[24] Buonicore WTDAJ. Air Pollution Engineering Manual, Air and Waste Management Association, Van Nostrand Reinhold, New York, 1992.

[25] Crocker BB. Air pollution control methods. In: Kirk-Othmer Encyclopedia of Chemical Technology. John Wiley & Sons, Inc, 2000.

[26] Pham TD, Lee BK. Feasibility of silver doped TiO2/Glass fiber photocatalyst under visible irradiation as an indoor air germicide. Int J Environ Res Public Health 2014;11:3271–88.

[27] Zhao J, Yang X. Photocatalytic oxidation for indoor air purification: a literature review. Building Environ 2003;38:645–54.

[28] Dezhi S, Sheng C, Chung JS, Xiaodong D, Zhibin Z. Photocatalytic degradation of toluene using a novel flow reactor with Fe-doped TiO2 catalyst on porous nickel sheets. Photochem Photobiol 2005;81:352–7.

[29] Park HW, Park Y, Kim W, Choi W. Surface modification of TiO2 photocatalyst for environmental applications. J Photochem Photobiol C: Photochem Rev 2013;15:1–20.

## 6

Air Pollution Monitoring and Prediction

Sheikh Saeed Ahmad, Rabail Urooj and
Muhammad Nawaz

Additional information is available at the end of the chapter

1. Introduction

One of the most important emerging environmental issues in Asian cities is air pollution. Air pollution is an atmospheric condition in which the concentration and duration of certain substances present in the air produce injurious and destructive effects on both man and the surrounding environment [1]. The most common pollutants in air are sulfur oxide, nitrogen dioxide, carbon monoxide and dioxide, and particulate matter.

Geographical Information Systems (GISs) are computer-based applications used for mapping and analyzing the earth and related spatially distributed phenomena. GIS applications integrate unique visualizations with common databases, which make it possible to capture, model, manipulate, retrieve, analyze, and present the geographically referenced data. Compared to other information systems, GIS systems have advantages, including the high power of analyzing spatial data and handling large spatial databases.

GIS applications can be used in air quality management and for controlling pollution, for handling and managing large amount of data. GIS systems manage spatial and statistical data, which facilitates depiction of the association between the frequency of human activities leading to bad environmental health and poor air quality. GIS modeling and statistical analysis also enables to examine and predict the impact of climatic variables on air pollution. In this way, GIS systems help in monitoring air pollution and emissions of pollutants from different sources.

Air pollution mapping is a helpful method for determining the concentration of pollutants. As the result of air pollution mapping, overviews of pollution in cities can be created and their sources of pollution emission can be identified, which help in controlling emissions. Different studies have been executed on air pollution in conjunction with GIS [2-11]. Consequently, GIS applications in air monitoring are necessary to determine air quality to reduce pollution to such a level at which harmful impacts on human health and the environment is reduced.

With the help of GIS applications, an output report of pollutants in Air Quality Management Systems (AQMSs) can be achieved in the form of three-dimensional (spatial) records. In AQMS emission time, concentration and place of air pollutants are regulated in order to achieve the predefined air quality standards of ambient air. It encompasses the estimation of the pollutants' emission schedule in a way to determine the consequences to air quality and the design of alternative programs for emission control in order to meet air quality standards, which are subject to some limitations, for example, technological viability and lowest charges. For environmental modeling with GIS applications, AQMSs are considered to locate monitoring stations, for development of geospatial model for air quality, and for spatial decision-support systems. However, the most significant step in an AQMS is data mining. The data mining method is a skill, which is used to analyze the data, uncover hidden patterns, and find interesting information from large amounts of data or huge databases. The most commonly used technique in data mining is artificial neural networks [12].

The human brain consists of a large number of neurons connected to each other by synapses to make networks, and these networks of neurons are called neural networks, or natural neural networks. Similarly, the artificial neural network (ANN) is basically a mathematical model of a natural neural network. The ANN uses a mathematical or computational model based on connectionist approach for solving the given problem. The concept of ANN is derived from biological neural network systems. The key applications of neural networks are control systems, classification systems, and prediction and vision systems.

Three basic components are important in order to make functional model, like: synapses of neuron; an added that sum all input in form of weights; and activation function. In Figure 1, synapses are shown by weights. Basically, a strong connection between input and neuron is noted by synapses or value of weight. Negative values reflect inhibitory connections, whereas excitatory connections are shown by positive values. Activation functions regulate the output of neurons within an acceptable range from -1 to 1.

1.1. Sources

Air pollution takes place due to natural and anthropogenic activities. But air pollution as the result of man-made activities like fossil fuel combustion, construction, mining, agriculture, and warfare are the most significant and cause problems in the atmosphere [13].

Basically, two types of pollution sources have been categorized, i.e., Stationary and Mobile. The stationary source is a type of source that is fixed or is a preset pollutant emitter, for example, fossil fuel burning power plants and refineries. The mobile source is a nonstationary type of pollutant emitter, for example, vehicles. The most emerging and leading cause of air pollution is the motor vehicle [14]. Pollutants that are emitted directly from the source into the air are known as primary pollutants, for example, carbon dioxide, carbon monoxide, sulfur dioxide, etc. When these primary pollutants react in atmosphere with each other to form another type of pollutants, they are called secondary pollutants, which are not directly emitted

Fixed input $x_0 = \pm 1$

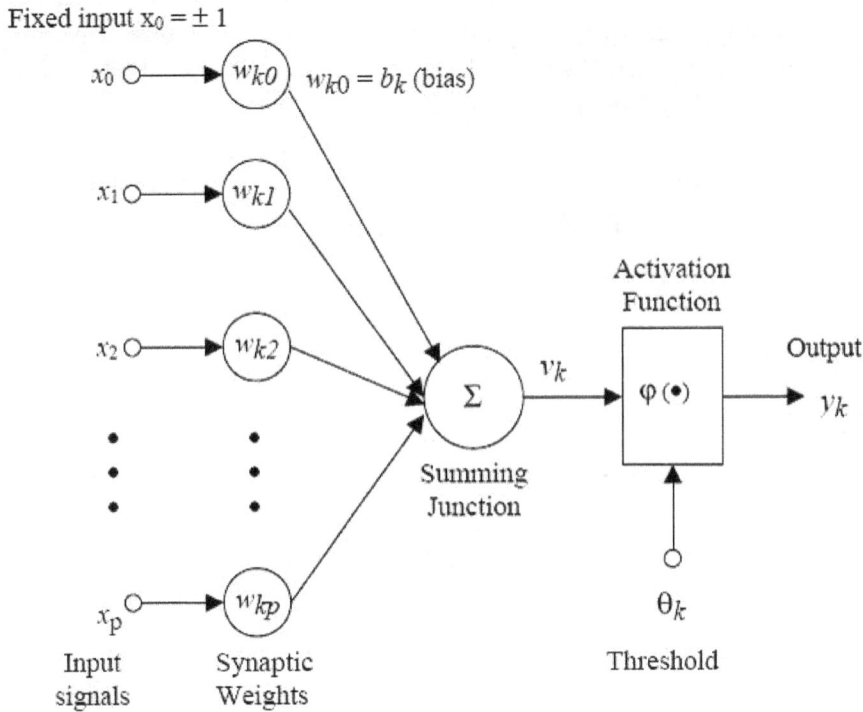

Figure 1. Model of a neuron

but formed as a result of primary pollutants' reaction in the atmosphere. For example, ozone forms when nitrogen oxides react with hydrocarbons in the presence of sunlight, and the resulting nitrogen dioxide reacts further with oxygen and forms ozone as pollutant.

1.2. Health effects

Air pollution and its effects in rural and urban areas are directly related to the ongoing activities. For example, in cities, pollution is related to the products of combustion in industries and vehicles. Many large cities all over the world exhibit excessive levels of air pollutants. Among all dangerous pollutants, nitrogen dioxide (NO_2) is important due to its capacity of causing dangerous effects on humans and the environment, which results in photochemical oxidation and acid rain.

The effects of air pollution cannot be ignored even within homes. Many air pollutants can cause cancer and other diseases among inhabitants. In 1985, it was reported that indoor toxic chemicals are three times more potent in causing cancer than outdoor air pollutants [15]. In America, health issues caused by buildings are called "sick building syndrome"[16].

1.3. Case study

In Pakistan, air pollution is emerging as a serious problem in its mega cities, which needs to be monitored and addressed at the root level in order to reduce the lethal impacts of pollutants on man and environmental health. The present study of Pakistan focuses on the most important twin cities of Pakistan, which are Rawalpindi and Islamabad. Both cities are commonly viewed as one unit and are 15 km apart. The study area with 135 sampling locations is shown in Figure 2. The climatic condition of Rawalpindi and Islamabad is sub-humid to tropical, with hot and long summers (May to August) accompanied by a monsoon season (July to August) followed by short and mild winters (October to March). The average low temperature is 12.05 °C in January and average high temperature is 31.13 °C in July.

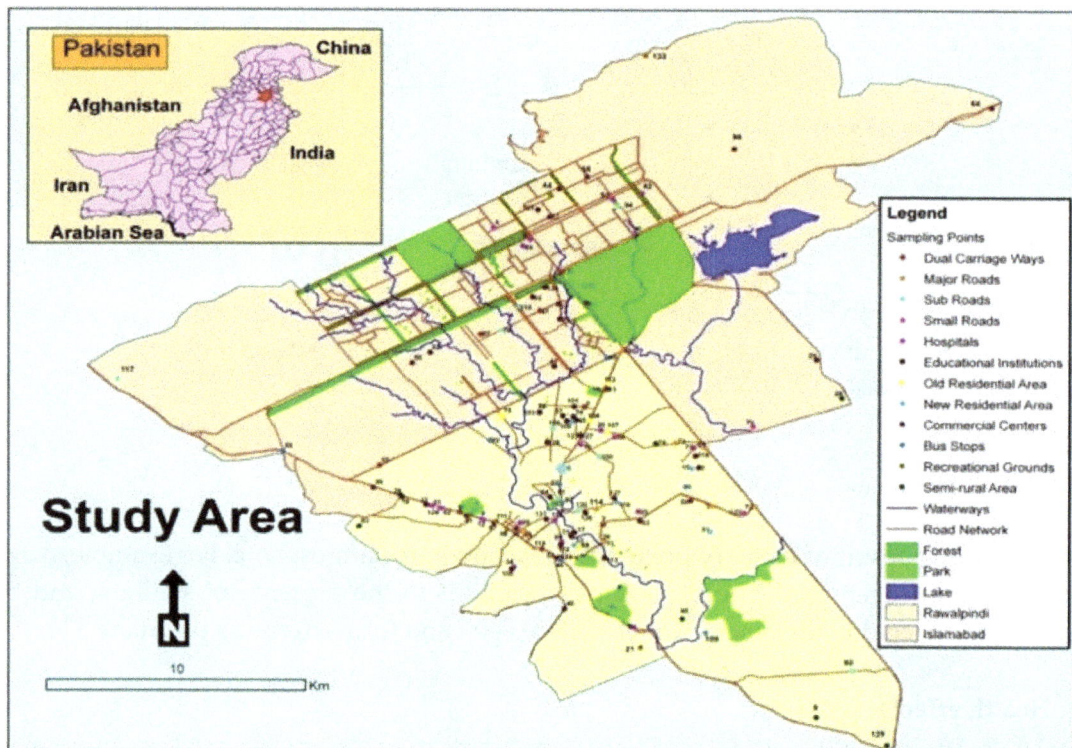

Figure 2. Base map

For the monitoring campaign, the maximum area (135 sampling sites) was covered in order to represent different traffic intensity and congestion levels in the urban area of Rawalpindi and Islamabad, for sampling. These sites included dual carriageways, major, linking, and small roads, healthcare centers, educational institutes, commercial areas, old residential areas, modern residential areas, recreational spots and semi-rural areas.

Research was carried out in order to monitor the NO_2 concentration in the ambient air of Rawalpindi city. Passive samplers were used within the city from January to December in 2008. The average concentration found was 27.46±0.32 ppb. The highest concentration was recorded near the main roads and in the vicinity of schools and colleges due to the large number of transport vehicles, which exceeded the set limit concentration value given by the World Health Organization.

2. Experimental design

2.1. Passive Sampling of NO_2

The most frequent method in monitoring studies for passive sampling of NO_2 is using diffusion tubes described by Atkins [17]. This method for NO_2 measurement is reliable, easy to handle, and it is an inexpensive method for screening air quality. Moreover, passive samplers are preferably appropriate for extensive spatial measurement of NO_2, and they have been reported in many studies of NO_2 monitoring of air in many countries like the United Kingdom, USA, France, Turkey, Argentina, and China [18].

Basically, passive samplers are designed on the principle of air diffusion having an efficient absorber at one end of the tube, and the flow rate (sampling rate) at constant temperature can be measured by using Flick's Law [19]. For that, the length and diameter of diffusion tubes are known, whereas sampling by using diffusion tubes is independent of air pressure.

2.2. Neural network design

From different sampling sites covering the whole study area, data was collected for neural network analysis. Collected data was fed to the neural network that has area_id, season_id, temperature, humidity, rainfall, and the respective concentrations as columns. For the neural network, the marked value was set to predict concentrations and rests were used as input to the neural network.

Neural network has two phases: training and testing. In the first phase (training), the network is trained by providing the complete information about the characteristics of data and observable outcomes to perform a particular task.

A neural network can develop a model that learns the relationship between input data and the desired outcome in the training phase. In the testing phase, testing data are provided as input. The performance of the testing phase depends upon the training phase (it depends on the number of samples that are provided during the training phase and also on the number of times that the network is accurately trained. However, it is impossible that the output is 100% precise for any network input. MS Access was used as the database engine because it is easy to use for all.

For testing the neural network, the cross validation method is used by using holdout method in which data was divided into testing and training data. The database consisted of two tables: training_ data and testing_data. The function of training_data is to train the ANN by adjusting weights in order to maximize the predictive ability of ANN and minimize error during forecasting. Testing data was used to test the prediction accuracy of ANN on new data. The structure of training data and testing data is given in Table 1.

In Table 1, the first key "id" is primary key, which contains the number that indicates row number and the second key "loc_id" contains the number that indicates location from where data is gathered, loc_name indicates the name of location and the next six fields indicate position of location with respect to north and east. The next two indicate temperature and humidity levels.

The 13th and 14th fields indicate concentration of NO_2 and level of concentration value. The last field of dataset contains week number, which indicates the number of weeks in which data is gathered from particular location. The attribute for testing data are the same in the testing data structure.

Field Name	Data type	Primary key	Field size
Id	Number	Yes	Long Integer
loc_id	Number		Long Integer
loc_name	Text		50
map_id	Number		Long Integer
north_d	Number		Long Integer
north_m	Number		Long Integer
north_s	Number		Long Integer
east_d	Number		Long Integer
east_m	Number		Long Integer
east-s	Number		Long Integer
Temp	Text		50
Humidity	Text		50
Concentration	Number		Long Integer
con_level	Number		Long Integer
Week	Number		Long Integer

Table 1. Structure of training data

For designing a network, we need to specify the architecture of a neural network by designing a number of hidden layers and units in each layer along properties of network that describe error function and network activation.

For optimal generalization of collected data, two types of architectures: the rtNEAT (real-time neuro evolution of augmented topologies) architecture with evolution algorithm and the feed forward architecture with back propagation algorithm of ANN are used in order to ensure high accuracy of ANN prediction about impacts of NO_2 concentration achieved in future. This rtNEAT architecture is used to train neural network with evolutionary algorithm, which has three steps, i.e., selection, mutation, and reinsertion. But before the training of neural network, the topology has to be created in the design of the neural network. A neural network is a connection of neurons, which contains three types of nodes: input, output, and hidden node. All nodes are randomly created during its execution.

Table 2 describes the properties of network, which contains an error function and network activation parameters. These properties are functional to all tested networks by the architecture search method and manually selected network.

Parameter	Value
Input activation FX	Logistic
Output name	Concentration
Output error FX	Sum-of-squares
Output activation FX	Logistic

Table 2. Network properties

The logistics function has a sigmoid curve and sum of squares. The sum of squares is the most frequent function error, which is used for the classification problem. The error is the sum of the square differences between the real input value and neural network target value.

2.3. Architecture search

A heuristic search is used to search the dataset for the best networks. Heuristic methods are used to speed up the process of finding a satisfactory solution. The architecture search for the designed neural network NO_2 is given in Table 3.

ID	Architecture	# of Weights	Fitness	Train Error	Validation ErrorError	Test ErrorError	AIC	Correlation	R-Squared
1	[8-1-1]	11	0.079965	11.371084	12.151164	12.505404	-4220.122343	0.652911	0.424949
2	[8-20-1]	201	0.080369	11.295746	12.093373	12.442678	-3846.484024	0.653839	0.427147
3	[8-12-1]	121	0.080841	11.017718	12.044772	12.37002	-4030.333807	0.668927	0.446734
4	[8-7-1]	71	0.080593	11.182193	12.108417	12.407984	-4116.153129	0.662147	0.438331
5	[8-16-1]	161	0.081507	10.941978	11.986108	12.26894	-3956.935279	0.670637	0.448777
6	[8-18-1]	181	0.080474	10.917611	12.026946	12.4264	-3919.068903	0.676823	0.457557
7	[8-14-1]	141	0.080839	11.105445	12.044827	12.370266	-3982.744071	0.666178	0.44366

Table 3. Heuristic architecture search for NO2

2.4. Training of neural network

The next step is to train the neural network for the NO_2 dataset by using the propagation algorithm. Weight change is calculated by the quick propagation algorithm by utilizing the quadratic function $f(x) = x^2$. In neural networks, several layers contain neurons in each layer that are connected with each other like neurons in the input layer connected to one or more neurons of the hidden layer, which are further connected to the output layer's neuron. With each presentation in neural network, error is computed as the difference between network output and observable output. The combination of randomly assigned weight (giving low error) replaces weights that are at the first location. This is called training to adjust the connection weights to enable the network to produce the expected output. Two different weights having two different error values are two points of a secant. Relating this secant to a quadratic function, it is possible to calculate its minimum $f'(x) = 0$. The x-coordinate of the minimum point is the new weight value.

$$S(t) = \frac{\partial E}{\mathbb{R}w_i(t)}$$

$$\Delta w_i(t) = \alpha \cdot \frac{\partial E}{\partial w_i(t)} \text{(Normal back propagation)}$$

$$\frac{\Delta w_i(t)}{\alpha} = \frac{\partial E}{\partial w_i(t)}$$

$$S(t) = \frac{\partial E}{\partial w_i(t)} = \frac{\Delta w_i(t)}{\alpha}$$

$$\Delta w_i(t) = \frac{S(t)}{S(t-1) - S(t)} \cdot \Delta w_i(t-1) \text{(Quick propagation)}$$

Here w =weight, i =neuron, E =error function, t =time (training step), α= learning rate, and μ= maximal weight change factor

The quick propagation coefficient was set to 1.75, learning rate was 0.1, and iterations were 500. The training graph for dataset errors for NO_2 is shown in Figure 3.

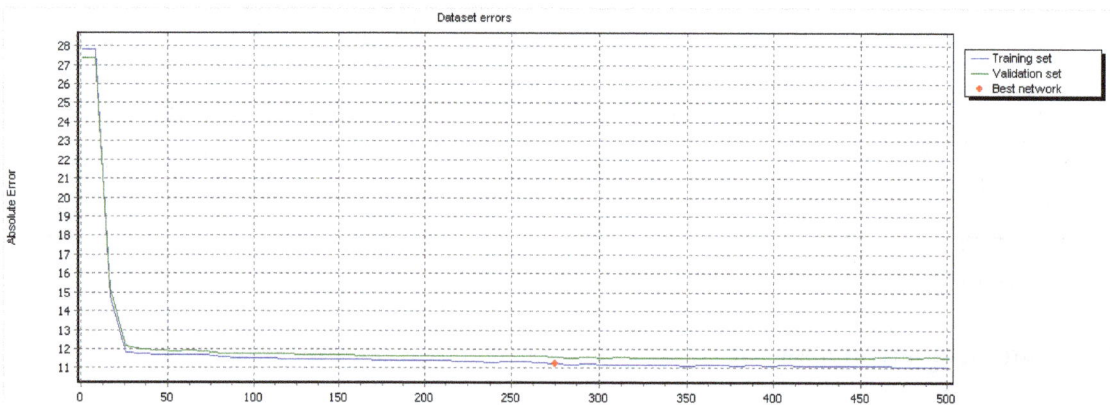

Figure 3. Dataset errors for the NO_2 dataset

The training graph of correlation for NO_2 is shown in Figure 4.

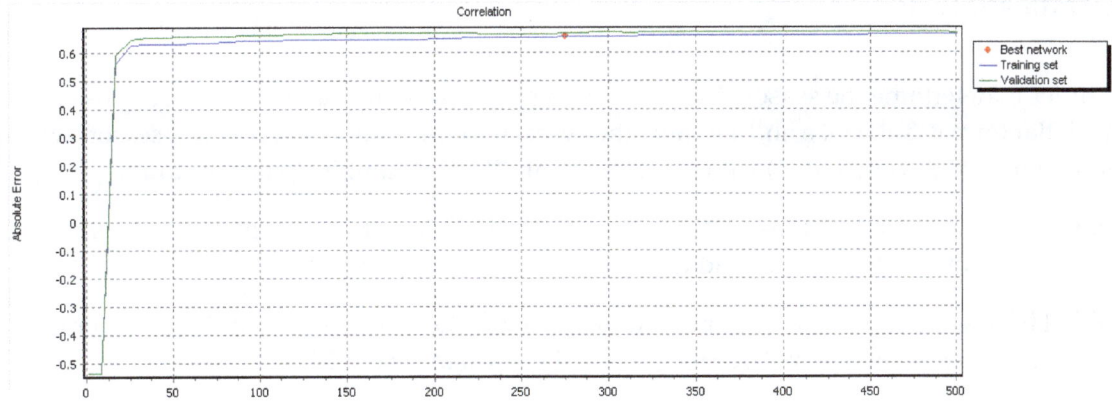

Figure 4. Graph of correlation for NO_2

The graph of error improvement – network errors for NO_2 is shown in Figure 5.

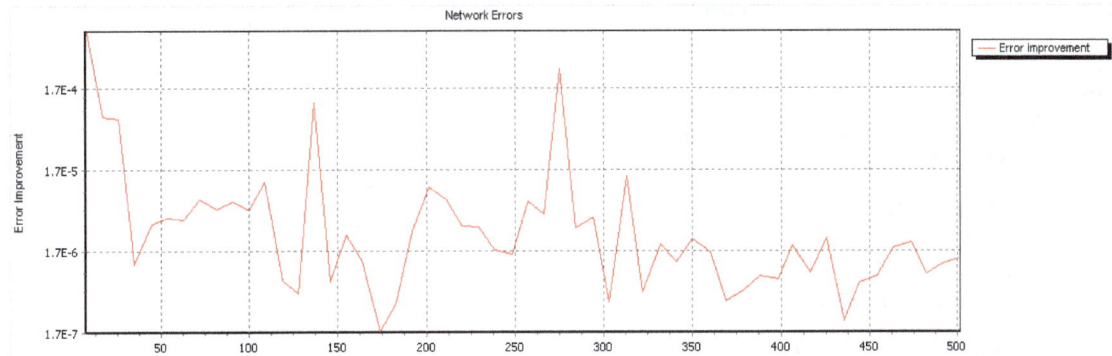

Figure 5. Network errors for NO_2

The error distribution of network statistics obtained after training of neural network is shown in (Figure 6).

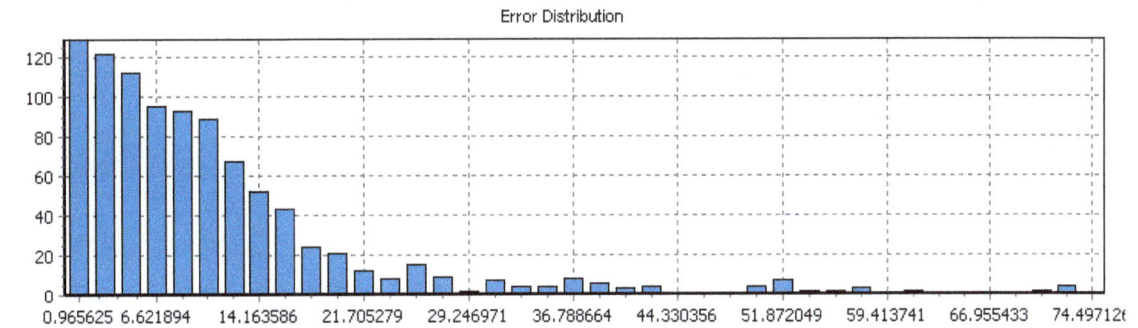

Figure 6. Error distribution for NO_2

3. Data analysis

In order to determine the seasonal variation and statistical significance, results are presented in tabular format. Tables 4 a and 4 b show the average concentration level of NO_2, season-wise, along standard deviation (SD) values measured at different sampling sites of study.

Table 4 a shows average values of NO_2 concentration in different seasons of 12 major sampling categories in urban Rawalpindi and Islamabad from November 2009 to July 2010.

Table 4 b shows the seasonal average concentration of NO_2 of 12 major sampling categories in urban Rawalpindi and Islamabad from September 2010 to March 2011.

Table 5 presents NO_2 concentration for each selected category, as described in study area profile, to understand the general trends of NO_2 concentration levels among different categories during the course of experimental period.

	Sampling Categories	Mild Winter (Nov)	Winter (Dec to Jan)	Early Spring (Feb)	Spring (Mar)	Mild Summer (April)	Summer (Pre-Monsoon) (May to June)	Monsoon (July to August)
	NO_2 Conc. (weekly basis)	(ppb)	(ppb)	(ppb)	(ppb)	(ppb)	(ppb)	(ppb)
Rawalpindi	Dual Carriage Ways (5)	87±19.78	98±26.87	63±12.29	53±6.49	44±10.64	22±4.22	18±1.91
	Major Roads (10)	60±12.19	68±9.56	52±13.52	45±10.23	36±8.97	26±5.88	19±4.74
	Sub-roads (6)	74±20.50	86±24.47	60±16.49	50±11.05	38±12.65	33±13.01	21±4.39
	Small Roads (3)	55±9.78	63±4.89	47±5.57	40±8.24	31±3.40	25±4.68	18±4.81
	Public Hospital (5)	48±18.71	63±18.40	37±0.74	29±2.29	22±2.24	18±0.79	14±0.96
	Private Hospitals (8)	61±14.47	75±14.19	38±1.16	32±2.0.3	25±2.29	20±5.57	14±3.98
	Public EI (11)	85±30.58	95±32.94	75±23.75	63±17.94	47±17.37	31±10.14	20±1.94
	Private EI (17)	55±9.71	66±9.54	45±4.56	43±9.65	38±10.89	26±4.54	18±3.18
	Old Residential Areas (5)	83±15.24	95±16.09	55±13.32	51±6.66	37±6.44	26±2.54	19±1.05
	Modern Residential Areas (5)	65±20.07	73±14.89	69±24.49	59±12.55	36±7.13	28±5.08	21±2.61
	Commercial Area (2)	75±0.83	82±17	61±6.69	51±7.11	36±4.29	21±6.20	18±4.78
	Bus Stops (9)	74±20.26	83±31.47	69±33.78	58±17	39±17.32	28±8.41	20±5.25
	Recreational Spots (9)	75±38.40	87±40.76	62±36.39	56±21.88	43±19.97	31±11.12	19±2.37

	Sampling Categories	Mild Winter (Nov)	Winter (Dec to Jan)	Early Spring (Feb)	Spring (Mar)	Mild Summer (April)	Summer (Pre-Monsoon) (May to June)	Monsoon (July to August)
	NO$_2$ Conc. (weekly basis)	(ppb)	(ppb)	(ppb)	(ppb)	(ppb)	(ppb)	(ppb)
Islamabad	Dual Carriage Ways (3)	84±28.73	95±33.64	66±23.78	57±12.31	45±16.69	24±5.98	19±4.16
	Major Roads (3)	50±3.72	60±2.04	40±0.81	32 ±2	26±4.42	21±2.97	15±2.16
	Sub-roads (4)	54±6.06	67±6.39	49±6.49	43±7.24	38±12.79	25±1.38	18±1.26
	Small Roads (3)	59±12.65	64±6.33	51±9.60	44±8.93	35±4.53	26±3.66	20±3.08
	Public Hospitals (3)	44±0.58	57±0.29	39±0.29	32±0.58	23±1.47	19±0.51	15±1.71
	Private Hospitals (1)	42	56	38	30	24	19	14
	Public EI (5)	53±13.34	64±9.32	46±9.30	39±10.76	34±14.19	25±6.01	18±1.28
	Private EI (6)	58±11.23	63±7.18	49±7.72	39±9.93	31±5.85	24±2.66	17±1.77
	Commercial Area (1)	61	68	57	50	35	25	16
	Bus Stops (12)	72±14.25	78±16.23	65±7.51	55±5.23	34±6.22	25±3.21	19±2.56
	Recreational Spots (2)	62±5.97	69±4.58	57±2.45	48±1.59	38±2.48	25±3.15	17±1.56
	Semi-Rural Areas (7)	46±8.98	59±5.64	42±6.41	33±7.87	31±5.29	24±3.22	18±3.19

(a)

	Sampling Categories	Mild Winter (Nov)	Winter (Dec to Jan)	Early Spring (Feb)	Spring (Mar)	Mild Summer (April)	Summer (Pre-Monsoon) (May to June)	Monsoon (July to August)
Rawalpindi	Dual Carriage Ways (5)	30±4.00	51±7.52	88±22.28	100±26.42	63±18.52	50±20.32	
	Major Roads (10)	27±6.57	49±9.72	61±10.33	68±9.34	48±3.76	37±3.79	
	Sub-roads (6)	32±7.31	53±12.97	74±23.42	87±26.60	58±13.69	39±6.36	
	Small Roads (3)	28±3.65	37±2.53	53±6.30	62±2.58	54±12.39	43±11.13	
	Public Hospital (5)	20±0.98	32±5.46	48±18.01	64±18.03	40±6.13	29±3.94	
	Private Hospitals (8)	23±3.90	38±7.19	60±15.37	73±14.03	40±4.05	31±1.95	
	Public EI (11)	44±16.98	81±36.87	86±31.78	96±34.20	73±21.26	63±18.23	
	Private EI (17)	31±4.85	42±6.10	55±8.91	66±9.82	45±7.08	35±5.90	

Sampling Categories	Mild Winter (Nov)	Winter (Dec to Jan)	Early Spring (Feb)	Spring (Mar)	Mild Summer (April)	Summer (Pre-Monsoon) (May to June)	Monsoon (July to August)
NO₂ Conc. (weekly basis)	(ppb)	(ppb)	(ppb)	(ppb)	(ppb)	(ppb)	(ppb)
Islamabad							
Dual Carriage Ways (3)	31±5.72	50±11.40	82±21.11	99±32.70	67±19.78	49±12.84	
Major Roads (3)	22±0.80	37±1.93	53±4.33	65±0.30	44±2.30	33±2.00	
Sub-Roads (4)	26±2.48	39±4.60	54±5.74	65±4.08	46±3.02	35±9.17	
Small Roads (3)	30±5.94	41±4.12	54±4.24	63±6.67	47±2.60	38±2.79	
Public Hospitals (3)	22±2.14	34±2.66	45±0.80	60±1.41	45±0.22	34±0.82	
Private Hospitals (1)	22	31	40	55	38	30	
Public EI (5)	31±7.41	40±3.60	52±7.94	64±8.39	46±9.34	37±9.62	
Private EI (6)	29±11.58	41±8.65	54±10.14	63±7.03	47±7.93	36±9.17	
Twin Cities							
Old Residential Areas (5)	27±2.97	61±14.74	84±14.18	95±16.51	58±12.41	48±10.06	
Modern Residential Areas (5)	32±7.86	49±11.70	66±20.07	75±16.16	60±19.16	48±16.53	
Commercial Area (3)	32±1.23	46±6.09	63±1.00	71±3.57	56±7.02	48±8.41	
Bus Stops (11)	32±9.11	53±20.30	76±20.07	87±32.40	69±31.34	54±19.54	
Recreational Spots (10)	37±18.55	52±25.23	71±37.63	84±39.83	57±29.71	46±24.78	
Semi-Rural Areas (7)	31±9.47	41±7.44	53±6.51	62±6.21	44±7.50	36±6.99	

(b)

Table 4. (a): Seasonal mean values of NO₂ from November 2009 to July 2010 (b): Seasonal mean values of NO₂ from September 2010 to March 2011

In Table 5 most of the sampling sites of study area showed nearly similar average concentration from month of November 2009 to March 2011. Maximum concentration of NO_2 shown on dual carriage ways.

The possible cause of such elevated levels of NO_2 concentration is extensive increase in number of vehicles, increase in population, busy roads, fuel inefficient vehicles, driving ways, and traffic jams. Gilbert reported that NO_2 is considerably related to both the distance from the nearest highway and the traffic count on the nearest highway [20].

The rest of the categories showed nearly the same average concentration. Major roads and sub-roads showed average NO_2 concentration levels of 53.56 ppb and 51.78 ppb, respectively. Sub-roads, bus stops, recreational spots, and educational institutions showed similar concentration levels of approx. 51 ppb.

Sampling Categories	No. of Sites	Average NO$_2$ Conc. (ppb)
Dual Carriage Ways	8	55.23
Major Roads	13	53.56
Sub-roads	10	51.78
Bus Stops	11	51.62
Educational Institutions	39	51.26
Recreational Spots	10	51.18
Old Residential Area	5	48.97
Small Roads	6	48.23
Commercial Area	3	47.59
Hospitals	17	47.44
Modern Residential Area	5	46.25
Semi-Rural Area	7	37.65

Table 5. Average NO$_2$ concentration levels in twin cities from November 2009 to March 2011

Educational institutions and recreational spots, being present close to the dual carriage ways, also experience elevated concentration levels. Old residential areas (48.97 ppb) showed slightly higher NO$_2$ concentration levels as compared to modern residential areas (47.59 ppb).

Narrow road, enclosing architecture, and congestion among the old residential areas result in traffic emission being trapped and buildup leading to higher NO$_2$ concentration levels, whereas in modern residential areas increased vehicular number is the major cause of elevated NO$_2$ levels. The minimum NO$_2$ concentration levels were indicated in semi-rural areas, that is 37.65 ppb. A study in Vilnius commented the same phenomena; NO$_2$ average rates depend upon traffic and are highest in cross roads and lowest at the background suburban areas [21].

For annual average concentration level of nitrogen dioxide, a spatial interpolation map has been developed by using inverse distance weighted (IDW). IDW in Figure 7 is clearly depicted as the areas of higher and lower concentration level of NO$_2$ in Rawalpindi and Islamabad.

Higher concentration levels are represented by darker shades while the lower concentration levels are shown with lighter shades. The maximum NO$_2$ values were found at the center of the city, where they reached the concentration of 83–110 ppb. Values were low on the outskirts of the city, with the lowest concentration in north (31–44 ppb).

A study in Vilnius commented the same phenomena; NO$_2$ average rates depend upon traffic and are highest in cross roads and lowest at the background suburban areas. Dual carriage ways, sub roads, major roads, commercial areas, old residential areas, and areas where schools and colleges are existing have higher concentration levels of NO$_2$. Intense traffic flow and congestion were the major reasons for these elevated levels of nitrogen dioxide concentration in those areas as vehicular emission is the predominant source of NO$_2$.

Vehicle growth rate in twin cities is extensively high. Load of traffic is continuously increasing with growing population rate and demand of motor producing industry. Due to this, traffic congestion is also increasing day by day with growing vehicle population, resulting in highest emission rates per vehicle.

The higher emission rate of NO_2 can also be attributed to the type of fuel and quality of fuel [22]. In Figure 7 Rawalpindi showed more concentration levels than Islamabad due their building patterns.

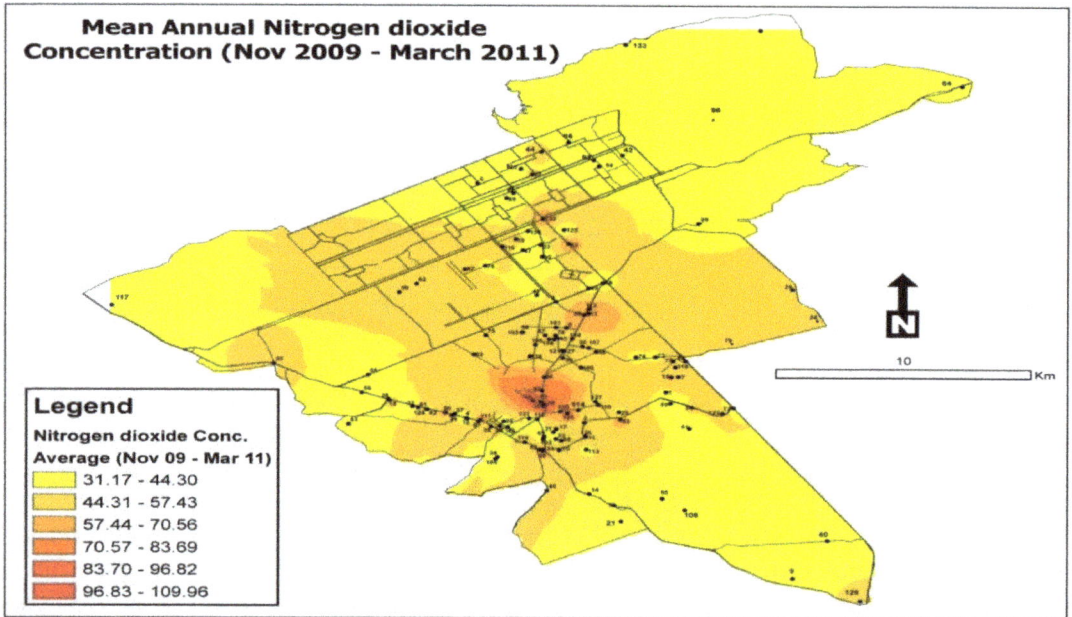

Figure 7. Spatial distribution of NO_2 concentration

3.1. Neural network data analysis

Based on the design of neural network, with the neural architecture and properties discussed, the data space is searched by using heuristic search method with 500 iterations and fitness criteria is set to Inverse Test error. The best top 5 networks explored from the space by the heuristic search are graphically shown (Figure 8).

Heuristic search is a problem-solving method that analytically searches a space of problem states. The best network is obtained when the absolute error gets minimum in the initial iterations so the best network out of the 5 best networks is shown (Figure 9).

Results for all data sets produced after training and testing data. Real vs. target graph represented a line graph of real- and network-predicted target values for record displayed in Table 6. X-axis shows the selected input column values and Y-axis represents network-predicted output values. Table 6 presents the summary of the real vs. output table after training.

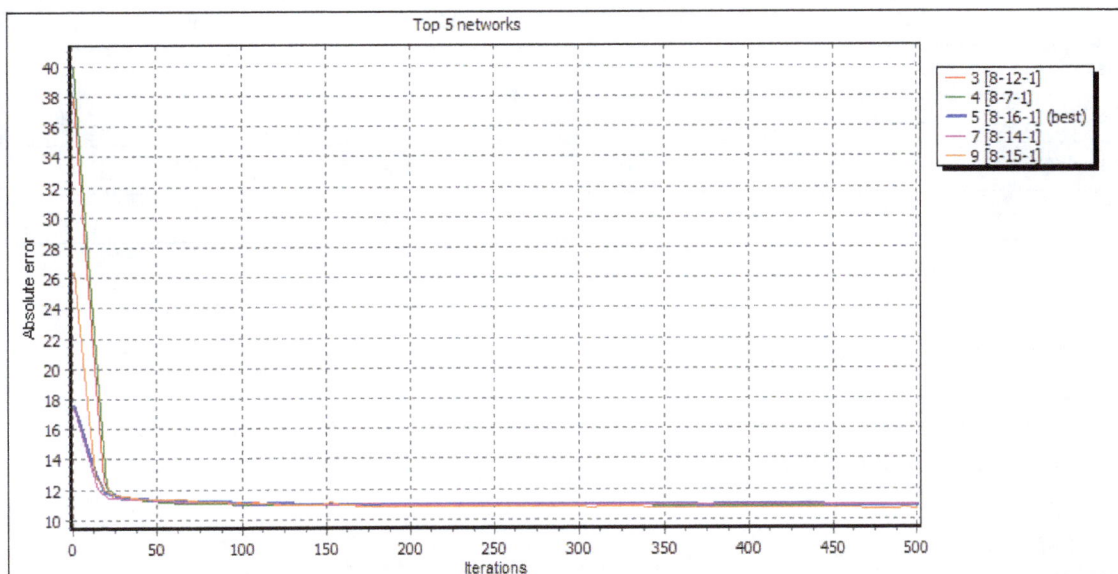

Figure 8. The top five networks explored by heuristic search approach

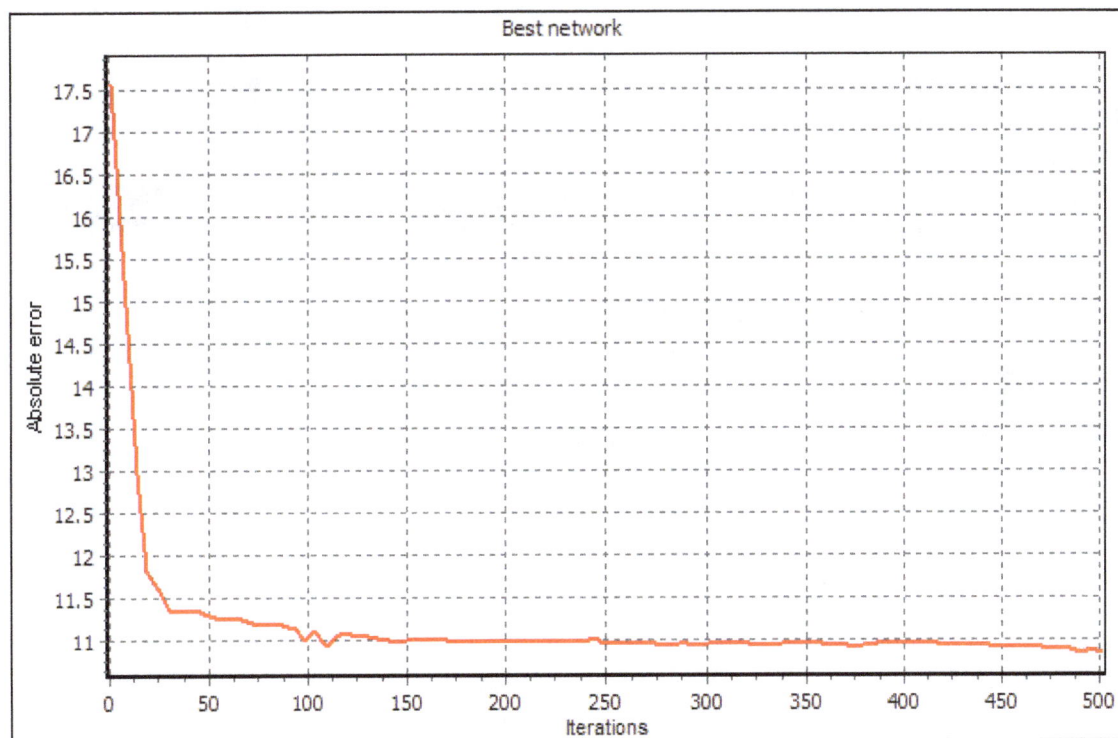

Figure 9. Network explored by heuristic search

	Target	Output	AE	ARE
Mean:	45.237265	45.09091	11.341221	0.250292
StdDev:	20.98552	13.97879	11.112997	0.180871
Min:	11.3	20.16986	0.004569	0.000171
Max:	132.72	63.353673	73.986765	1.096446

Correlation, 0.653989; R-squared, -0.290243

Table 6. Summary of real vs. target

The visualization for real vs. output with row number on x-axis and target/output (area_id) on y-axis is shown (Figure 10).

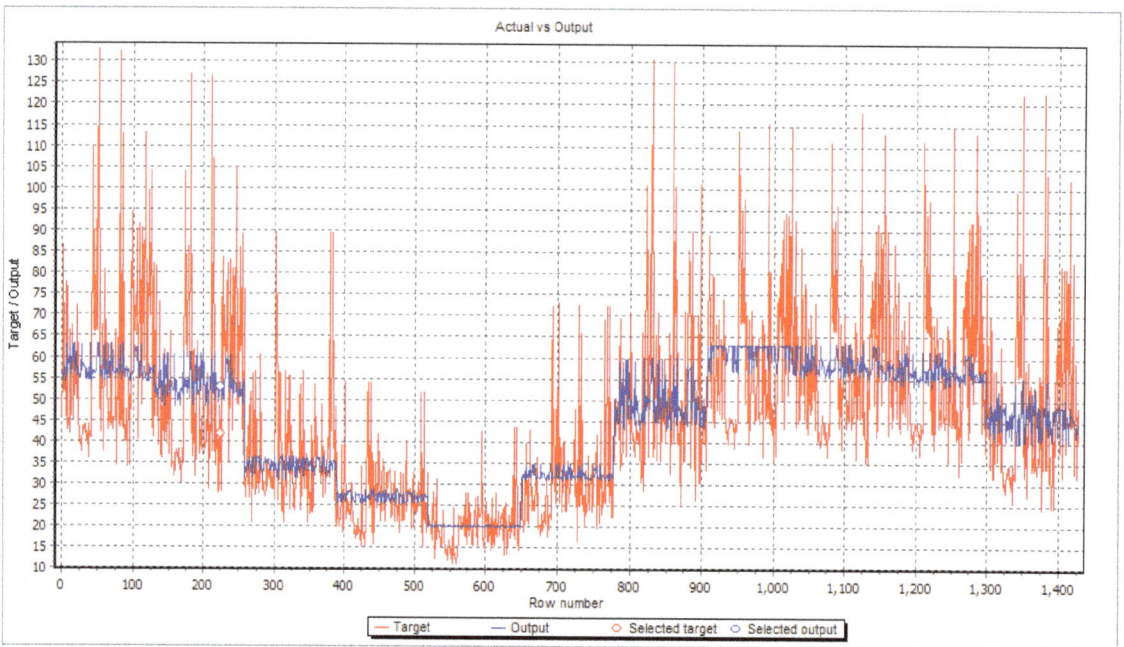

Figure 10. Real vs. network output

Figure 11 shows a scatter plot of the real and forecasted output values. X- axis presents the real values and Y-axis shows predicted network values.

Graph in Figure 12 shows the Network Error Dependence on values, which are numerically input in columns of data sheet. Through graph of Error Dependence, the ranges of the selected input column that can produce network error can be identified.

The last phase after the neural network is trained and tested is to query the network. The concentration is the output value for the neural network. So the input queries are subjected area_id, season_id, temperature, relative humidity, and rainfall (Figure 13).

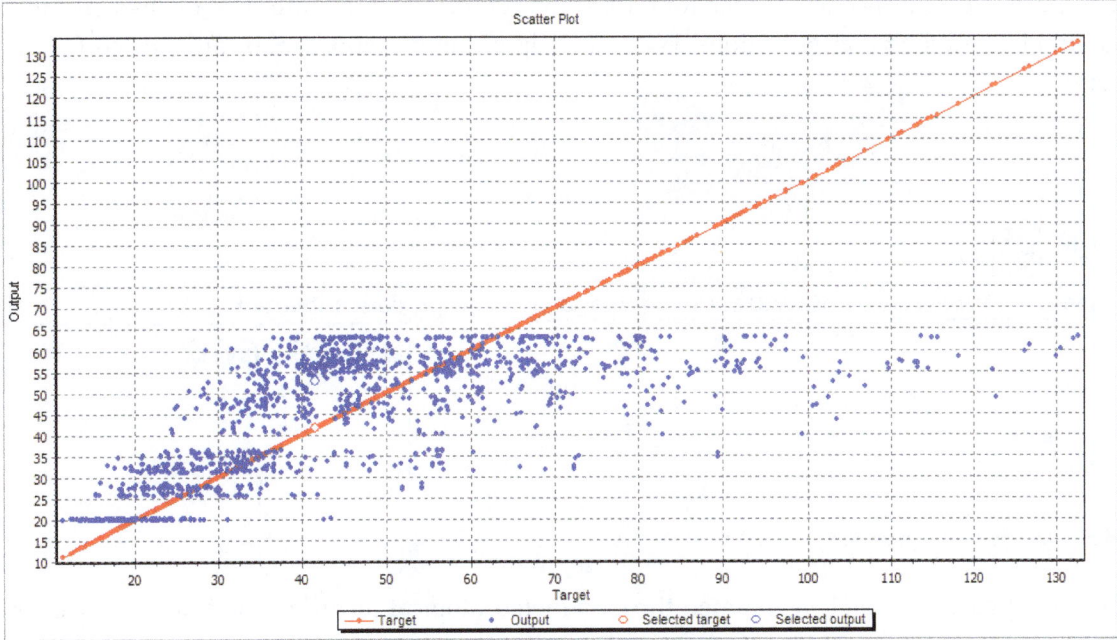

Figure 11. Scattered plot of real and output network values

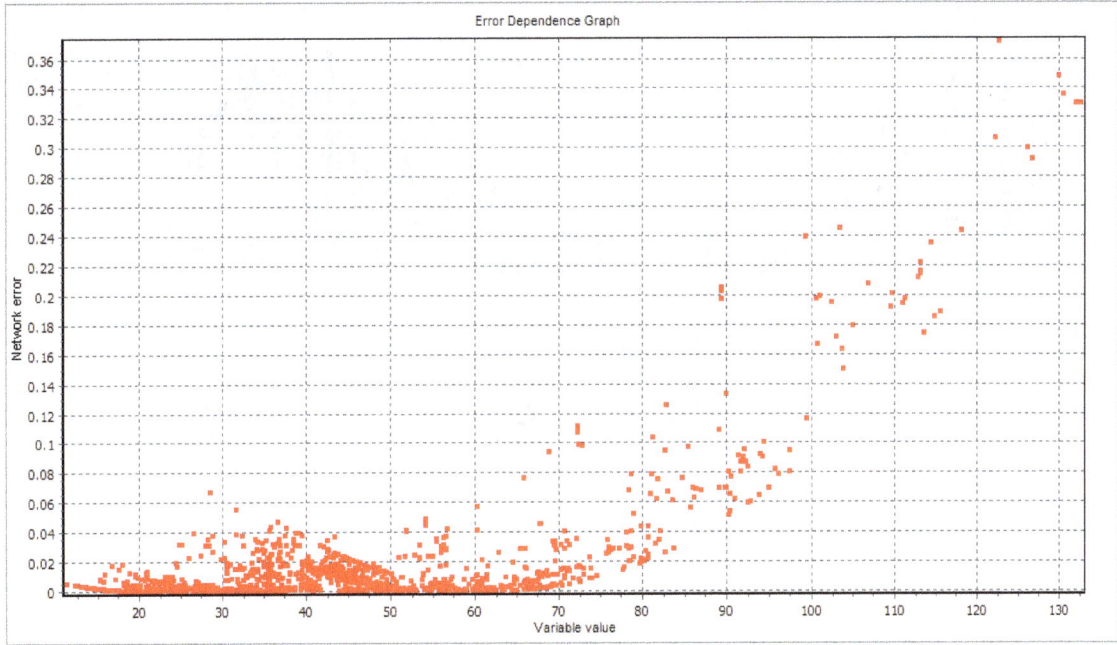

Figure 12. Graph of error dependence

The input Excel sheets are prepared for the GIS mapping. Sheets include area_id, their latitude, longitude, and their concentrations. With the help of interpolation, maps are created for the service.

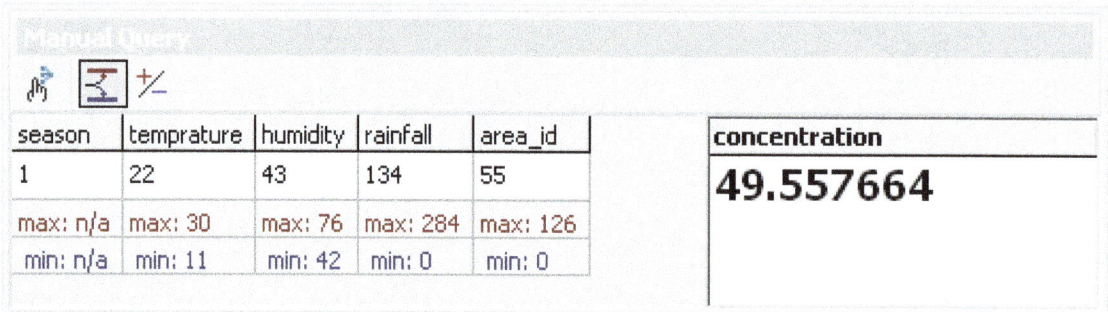

Figure 13. Excel sheet presenting manual query

Temporal variation can be explained through meteorological recorded conditions. However, most of the variations on a local scale are due to the impact of air pollutants.

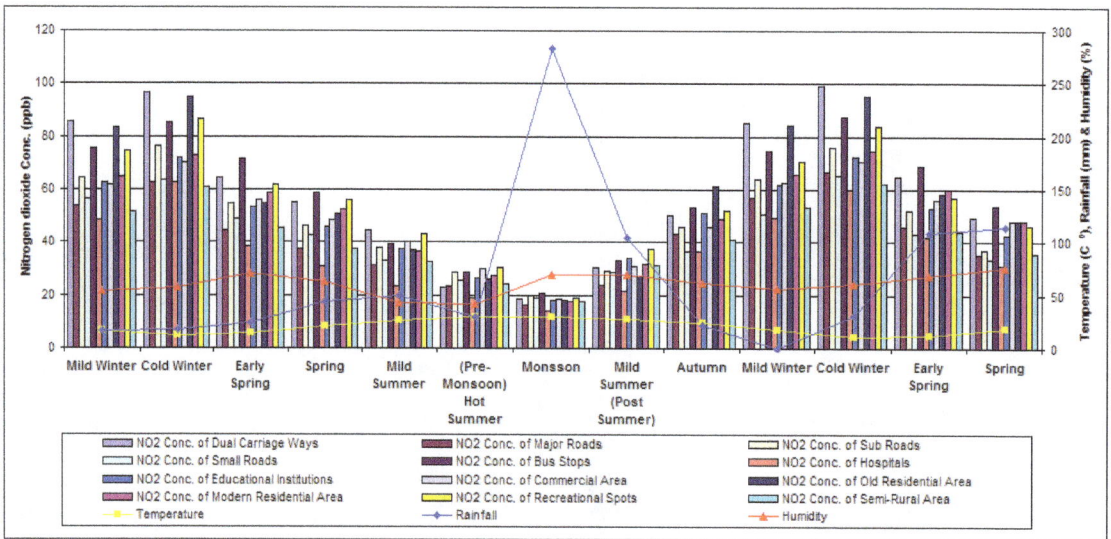

Figure 14. Relationship of rainfall, temperature, and humidity with NO_2 concentration (November 2009–March 2011)

Figure 14 indicates the positive association of NO_2 concentration level with humidity (RH in %) and negative association with the temperature. Figure 15 shows the concentration of NO_2 during summer when recorded temperature, rainfall, humidity are 31^0C, 67, and 17mm, respectively.

Figure 16 shows the concentration of NO_2 during the winter season at 11 ^0C, 68% humidity, and 9mm rainfall.

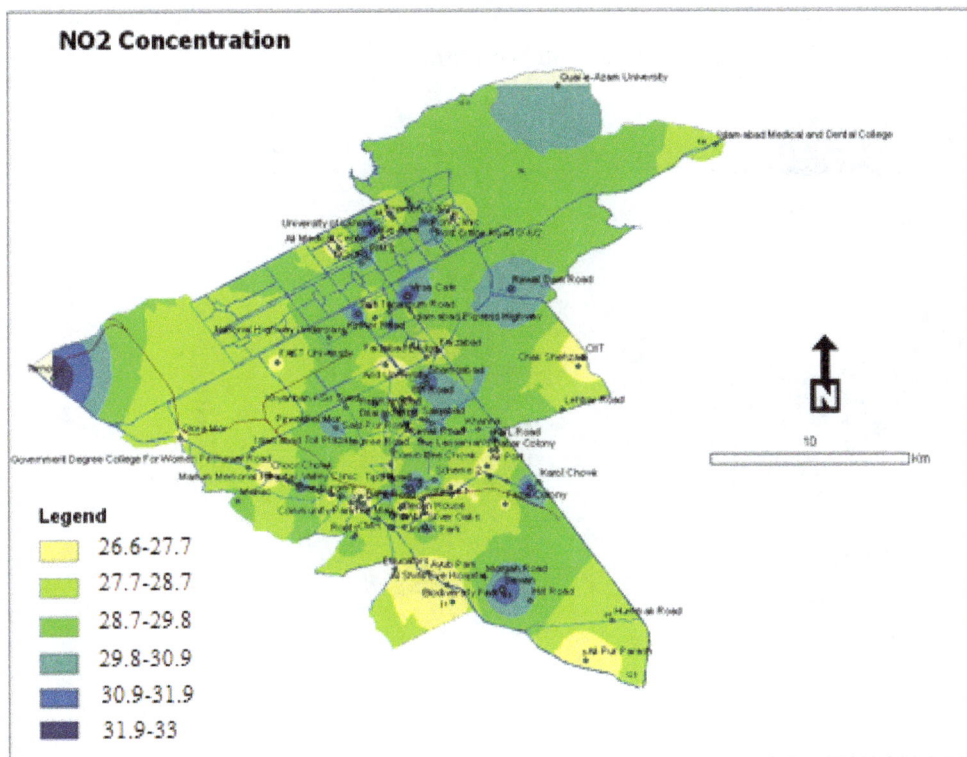

Figure 15. NO$_2$ concentration in summer

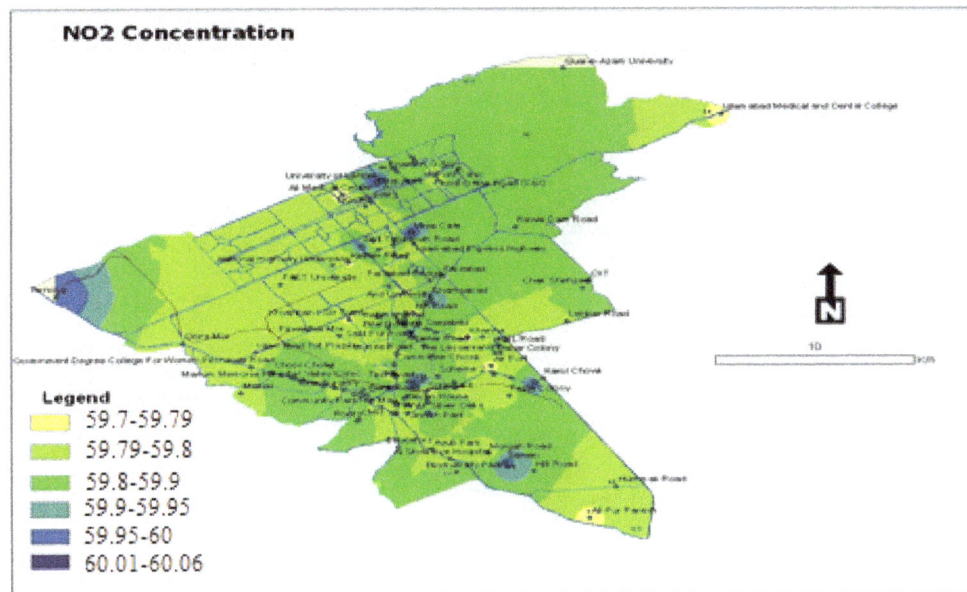

Figure 16. NO$_2$ concentration in winter

Concentration of NO_2 during the spring season, shown in Figure 17, when recorded temperature is 35°C, humidity is 58%, and rainfall is 60 mm.

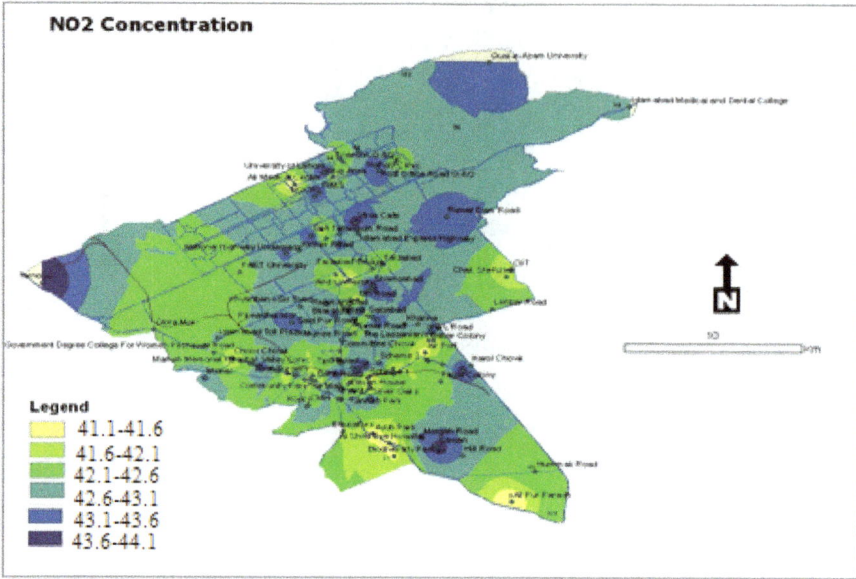

Figure 17. NO_2 concentration in spring

Figure 18 shows predicted concentration of NO_2 in autumn season when recorded temperature, humidity, and rainfall are 29 °C, 69, and 22 mm, respectively.

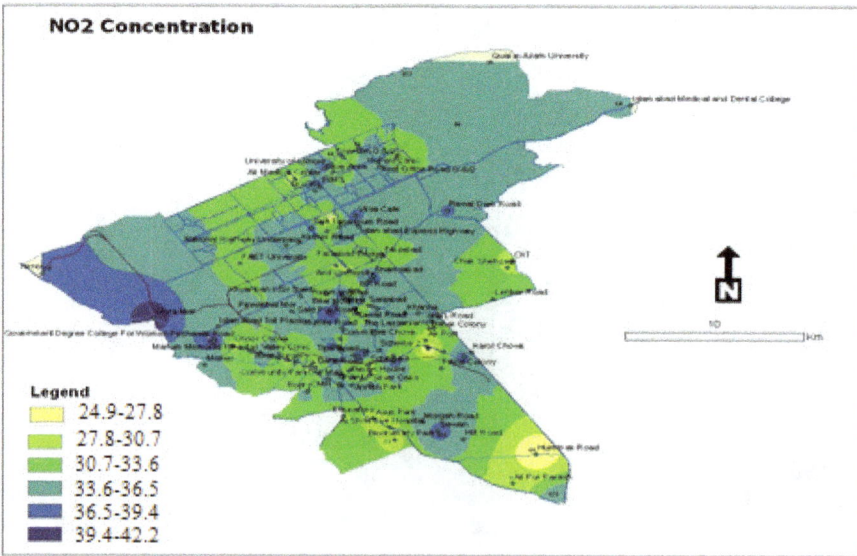

Figure 18. NO_2 concentration in autumn

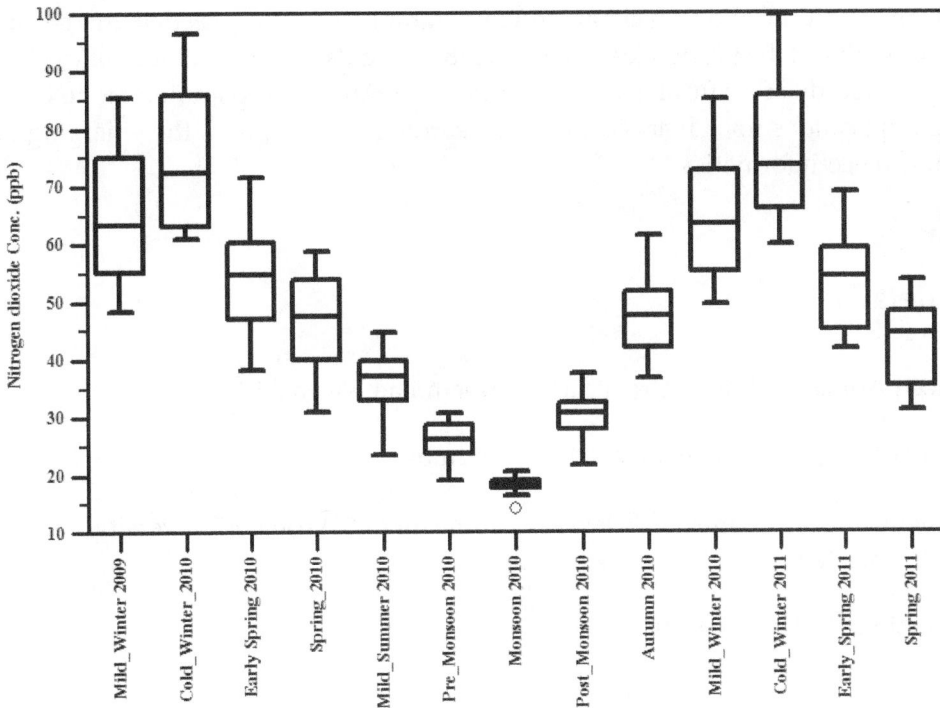

Figure 19. Seasonal variation in NO$_2$ concentration levels (November 2009 – March 2011)

Figure 19 shows that concentration of NO2 varies in different seasons. The months from May to August were months in which the minimum value of NO2 was recorded, and the maximum concentration was measured in the winter season from December to January.

4. Conclusion

NO$_2$ concentration levels were recorded on hourly and weekly basis in Rawalpindi and Islamabad city by using diffusion tubes. Artificial neural networks were trained to generalize the process of air pollutant spread over three dimensions. Prediction capabilities of ANN were analyzed through generalization by using hold-out evaluation method of classification. Results showed the advantage of using rtNEAT-like architecture of ANN where a neural network can modify its architecture to reduce the error up to the maximum possible limit. Results showed that annual average concentration of NO$_2$ concentration was 44 ± 6 ppb. However, the highest concentration was recorded in winter season near the dual carriage ways, schools, and colleges because of the higher number of transport vehicles on the road. This endorsed the fact that the reduced photolysis leads to the accumulation of NO$_2$ during winter due to less solar radiation. This is again attributed by the results of correlation, which reveal the negative correlation of nitrogen dioxide concentration levels with rainfall and temperature and the positive correlation with humidity. Moreover, the results of correlation reveal that the measured NO_2 concentration levels at different sampling areas exceeded the set limit of concentration value of the World Health Organization and Pak-EPA standard policy. This type of investigative study of artificial neural networks in the area of air pollution modeling shows promising applications for advanced machine learning algorithms in the emerging area of research called eco-informatics.

Author details

Sheikh Saeed Ahmad[1,2*], Rabail Urooj[1,2] and Muhammad Nawaz[1,2]

*Address all correspondence to: drsaeed@fjwu.edu.pk

1 Department of Environmental Sciences, Fatima Jinnah Women University, Mall Road, Rawalpindi, Pakistan

2 BZ University, Multan, Pakistan

References

[1] Mulaku C. Mapping and analysis of air pollution in Nairobi, Kenya. International conference on spatial information for sustainable development, Kenya, 2001.

[2] Gualtieri G and Tartaglia M. Predicting urban traffic air pollution: a GIS framework. Transportation Research – D; 1998; 3(5): 329–336.

[3] Pummakarnchana O, Tripathi N and Dutta J. Air pollution monitoring and GIS modeling: a new use of nanotechnology based solid state gas sensors. Science and Technology of Advanced Materials 2005; 6: 251–255.

[4] Afshar H and Delavar MR. GIS-based air pollution modeling in Tehran. Environmental Informatics 2007; 5: 557–566.

[5] Barnes J, Parsons B and Salter L. GIS Mapping of nitrogen dioxide diffusion tube monitoring in Cornwall, UK. Air Pollution 2005; 13: 157–166.

[6] Elbir T, Mangir N, Kara M, Simsir S, Eren T and Ozdemir S. Development of a GIS-based decision support system for urban air quality management in city of Istanbul. Atmospheric Environment 2010; 44: 441–454.

[7] Veen AVD, Briggs DJ, Collins S, Elliott S, Fischer P, Kingham S, Lebret E, Pryl K, Reeuwijk HV and Smallbone K. Mapping urban air pollution using GIS: a regression-based approach. International Journal of Geographical Information Science 2010; 11(7): 699–718.

[8] Vienneau D, de Hoogh K and Briggs D. A GIS-based method for modeling air pollution exposures across Europe. Science of the Total Environment 2009; 408: 255–266.

[9] Banja M, Como E, Murtaj B and Zotaj A. Mapping air pollution in urban Tirana area using GIS. International Conference SDI, Skopje, 15–17 September 2010, 105–114.

[10] Jensen SS. Mapping human exposure to traffic air pollution using GIS. Journal of Hazardous Material 1998; 61(3): 385–392.

[11] Kim JJ, Smorodinsky S, Lipsett M, Singer BC, Hodgson AT and Ostro B. Traffic-related air pollution near busy roads. American Journal of Respiratory and Critical Care Medicine 2004; 170(5): 520–526.

[12] Alexander SM. Data mining 2005. hhp//www.eco.utexas.edu/~norman/BUS, FOR/ course.mat/Alex/ (Accessed 12th February 2008).

[13] United Nation. Conference on the Human Environment, Sewden, 1972.

[14] United Nation. Environmental Performance Annual Report, New York, 2001.

[15] US-EPA. Health affects of different air quality index (AQI) levels caused by nitrogen dioxide, 2008.

[16] Miller D. Potential hazards of future volcanic eruptions. California, 1989.

[17] Atkins DHF, Sandallas J, Law DV, Hough AM and Stevenson K. The measurement of nitrogen dioxide in the outdoor environment using passive diffusion tube samplers. AEA Technology, 1986.

[18] Varshney CK and Singh AP. Passive samplers for NO_x monitoring: a critical review. The Environmentalist 2003; 23: 127–136.

[19] Palmes ED, Gunnison AF, Dimattio J, and Tomczyk C. Personal sampler for nitrogen dioxide. American Industrial Hygiene Association Journal 1976; 37: 570–577.

[20] Gilbert NL, Goldberg MS, Beckerman B, Brook JR and Jerrett M. Assessing spatial variability of ambient nitrogen dioxide in Montreal, Canada, with a land-use regression model. Journal of the Air & Waste Management Association 2005; 65: 1059–1063.

[21] Lozano A, Usero J, Vanderlinden E, Raez J, Contreras J, Navarrete B and Bakouri HEI. Air quality monitoring network design to control nitrogen dioxide and ozone applied in Granada Spain. Ozone: Science & Engineering 2011; 33(1): 80–89.

[22] Heywood JB. Internal combustion engine fundamentals. McGraw-Hill, New York, 1998.

7

Indoor air Quality Improvement Using Atmospheric Plasma

Kazuo Shimizu

Additional information is available at the end of the chapter

1. Introduction

Air pollution is serious problem for human health and for the environment including causing global warming. An interdisciplinary collection of new studies and findings regarding air pollution was previously published [1]. Governments around the world are managing air quality in their countries for the health of their citizens. The management of air pollution involves understanding sources of air pollution, monitoring contaminants, modeling air quality, performing laboratory experiments, controlling indoor air pollution, and eliminating contaminants through various methods. Research activities are carried out on every aspect of air pollution and its control throughout the world to respond to public concerns.

Another book raised concerns about environmental air pollution including indoor air pollution as well. In recent years, the deterioration of indoor air quality (IAQ) has become a concerning issue [2]. Modern construction methods make the spaces inside buildings more air tight. And, various harmful chemicals are used in building materials. Among the pollutants are volatile organic compounds contained in coated materials and adhesives. While activated charcoal filters used in homes and buildings can remove organic molecules, they have a limited lifetime and they must be replaced regularly to maintain good performance.

Another pollutant is particulate matter ranging from sub-micron to few micron meters including dust pollution caused by cigarette smoke, house dusts, and various fungi caused by hot and humid air. Air filters have been used to improve indoor air quality for many years. However, most filters used in homes have poor collection efficiencies for smaller particles (less than 10 micrometer in diameter). This low collection efficiency is a problem, especially for 2nd hand smoke, dust and pollen particles [3-5].

Living microorganisms such as bacteria and viruses also contaminate the indoor air. They may bring on respiratory disorders, asthma, or influenza. And, activated charcoal filters can incubate bacteria. So the other sterilization method is required for living microorganisms or virus.

Finally, while sterilization devices for indoor air may be commonly used in medical facilities (hospitals, clinics, etc.), these devices are uncommon for use in homes and general office buildings. While these issues could be eliminated by natural air flow, this method is energy inefficient and many houses cannot use this method. Because we spend most of our time indoors, indoor air quality is an important factor for healthy and comfortable lives.

From these points of view, there is a need for reliable devices for the home that can remove organic pollution and that can sterilize indoor air. Devices using atmospheric pressure plasma technologies, especially microplasmas, are very promising.

This chapter deals with improving indoor air quality using atmospheric plasma treatment [6]. Discussed are the results of a series of experiments on particulate matter (PM) precipitation and removal, odor control targeting ammonia and sterilization of *E. coli*. While such experiments are often performed in a small experimental chamber, these experiments were carried out in the relatively large space (23.4 m^3) shown in Fig. 1. These results reveal the performance of commercially available air treatment devices.

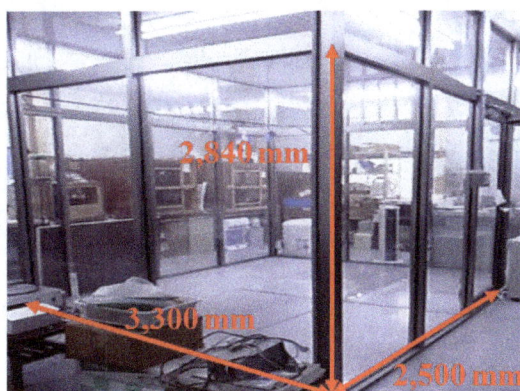

Figure 1. The experimental room for measuring indoor air quality improved by atmospheric plasma.

2. Experimental set up for PM precipitation

2.1. About particulate matter in the indoor air environment

Indoor particulate matter is a mixture of substances including carbon (soot) emitted by combustion sources, tiny liquid or solid particles in aerosols, fungal spores, pollen, and toxins present in bacteria (endotoxins). In many homes, most of the airborne particulate matter comes from the outside. However, some homes do have significant sources of indoor particulate matter that comes from various sources: cigarette smoking is the greatest single source of particulate matter in homes and buildings where people smoke; cooking: especially frying and sautéing; combustion appliances: for example, furnaces without a proper air filter; non-vented combustion appliances like gas stoves; wood-burning appliances like wood stoves and fireplaces: especially if the smoke leaks or *backdrafts* into the home; and mold growth [7].

Since indoor air contains various types of particulate matter, improving air quality requires a combination of high performance filters and active treating devices such as electrostatic precipitators and atmospheric plasma devices [8, 9].

2.2. Electrodes for particulate matter collection

The reactor shown in Fig. 2 consisted of a corona discharge needle type electrode system, a ground electrode, and a mesh filter electrode for collecting particulate matter. The needle type electrode system shown in Fig. 2 consisted of 5 parallel wires to which 4 needle type electrodes were attached to each wire. The diameter of the electrode assembly was 55 mm. The ground electrode consisted of parallel wires positioned at a discharge gap of about 5mm from needle type electrodes. At the outlet, a mesh filter placed 5mm from grounded electrode captured particles.

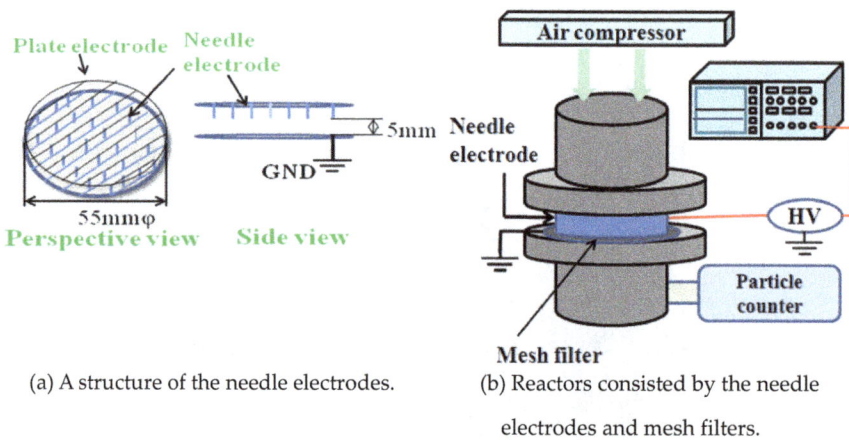

(a) A structure of the needle electrodes.   (b) Reactors consisted by the needle electrodes and mesh filters.

Figure 2. Schematic diagram of the needle electrodes and the reactor system with the mesh filters.

Figure 3. Mesh filter used for collecting dust. (x100)

Mesh filters like the one in Fig. 3 were used in this study. One side of the filters were coated by metal. The characteristics of the mesh filters are given in Table 1.

Aperture size [mm]	0.1
Wire diameter [mm]	0.06
Mesh count [mesh /inch]	150

Table 1. Characteristic of mesh filters.

We used the large, experimental room in Fig. 1 to carry out our indoor air quality experiments using the reactor in Fig. 2. We measured particulate matter collection, and ammonia removal from the indoor air. The fan in the corner of the experimental room shown in Fig. 4 circulated indoor air at a flow rate of 4.6 m³ / min.

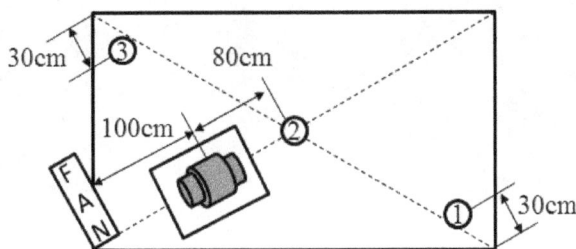

Figure 4. A reactor and a fan arrangement in the experimental room.

The reactor was placed 30 cm above the floor and 100 cm from fan. The gas flow through the reactor was about 112 L / min. The reactor was driven by a negative DC voltage of 5 kV.

To evaluate the collection of particulate matter, we took particle measurements at three locations #1, #2 and #3 in Fig. 4. We measured the number of particles using an optical particle counter (SHIMADZU, MODEL 3886). The diameter of the measured particles was 0.5 μm. During the measurement time of 2 hours, the relative humidity was 67% and the temperature was 23.7 C.

2.3. Microplasma generation for indoor air treatment

The microplasma electrode has been described in previous publications [6, 10]. In this study, the volume of the experimental room shown in Fig. 1 (23.4 m³) was much larger that the experimental chambers used in previous studies. So the microplasma electrode in Fig. 5 was enlarged corresponding to the area.

The microplasma electrodes consisted of a high voltage electrode and a grounded electrode separated by a discharge gap in the range 30 to 100 m. Both electrodes were coated with dielectric materials. When an AC voltage in the range from several hundred volts to 1 kV was

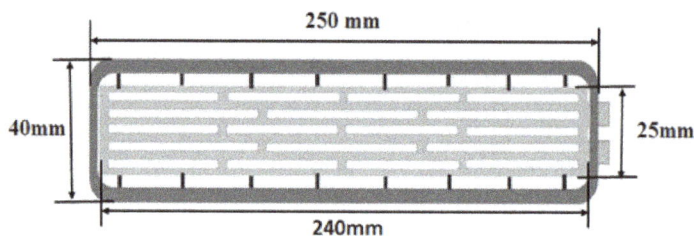

Figure 5. Microplasma electrodes utilized in this study.

applied to the high voltage electrode, numerous streamers were generated between the electrodes as shown in fig. 6 [11].

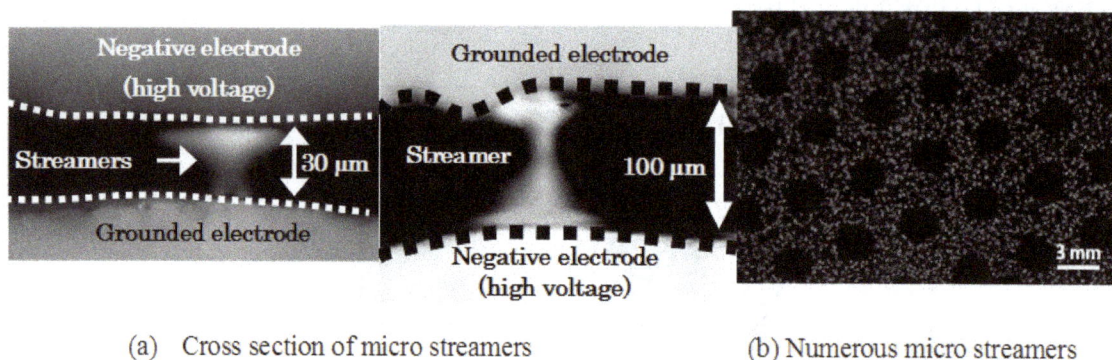

(a)   Cross section of micro streamers   (b) Numerous micro streamers

Figure 6. Generation of micro streamers (a); cross section, (b) front view.

The basic characteristics of a microplasma are similar to that of a typical dielectric barrier discharge (DBD) that typically has a discharge gap of 1mm or more. However, the diameters of streamers in microplasmas were observed to be in the range 10-20 um, which is narrower than that of DBD streamers [11]. The numerous microplasma streamers generate various active species, such as radicals, ions, and ozone that enhance the chemical reactions in the gas phase describing the next section.

3. Results and discussion

3.1. PM collection by the needle electrodes with the mesh filter

The particle concentrations in Fig. 7 were measured in the experimental room while applying a negative DC voltage of 5 kV.

The measured particle concentrations at the 3 measurement points shown in Fig. 4 were consistent indicating that the air in the experimental room was well mixed by the fan.

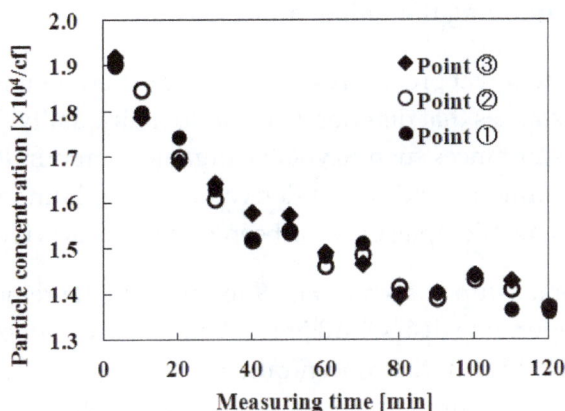

Figure 7. Particle concentration versus treatment time at 3 measurement points in the experimental room.

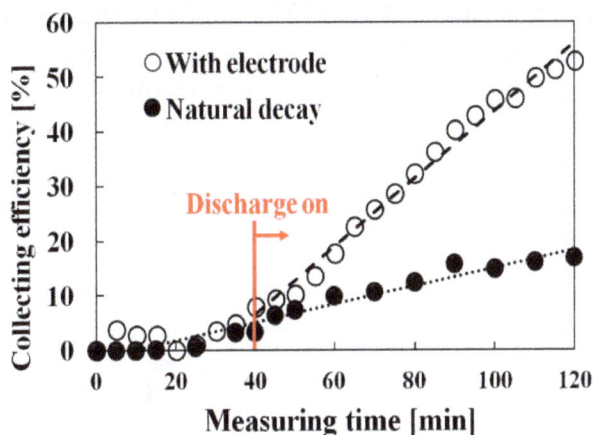

Figure 8. Decrease of the particles in the experimental room measured at position #2.

Airborne particles will deposit naturally on surfaces resulting in the natural decay of the particle concentration shown in Fig.8. When the power supply of the reactor was turned on for 40 minutes, the deposition of particles on the mesh filter within the reactor increased the decay rate compared to the natural decay (without any discharge). The collection efficiency of particle collection is calculated using equation (1) [12, 13].

$$\eta = \left(1 - \frac{N_A}{N_B}\right) \times 100(\%) \tag{1}$$

Particulate matter (PM) charged by the needle electrode was then trapped by the mesh filters. The effect of mesh filters and the discharge voltage, the number of filters and the flow rate have been previously published [14].

3.2. Odor removal for improving the indoor air quality

This section presents the removal process results from the experimental room (23.4 m³) shown in Fig. 1 of chemical substances that deteriorate the indoor air quality. The removal processes target various chemical substances such as volatile organic compounds (VOCs) derived from the building materials, ammonia and others derived from the human being and animals as well. Treatment of VOCs by microplasmas has been published previously [6, 15].

We used the experimental setup shown in Fig. 9 to measure the removal of ammonia from indoor room air. The evaporating dish with 3.5ml of liquid phase ammonia (25%) diffused into the air forming a relatively high initial concentration of ammonia. After confirming the initial concentration of the gas phase, ammonia reached a constant value, the experiment room was ventilated, to evaporate the ammonia again for 10 minutes. Using this procedure, the initial concentration of ammonia was set to 25ppm, which is an acceptable concentration according to the guideline values. Ammonia concentration changes in the experimental room were measured for 2 hours with a negative voltage of 5 kV applied to the needle. The needle electrode was placed in front the fan shown in Fig. 4 at a distance of 100 cm. To enhance the ammonia removal reaction, a fine water mist (3 μm) was added by an ultrasonic wave humidifier [16, 17].

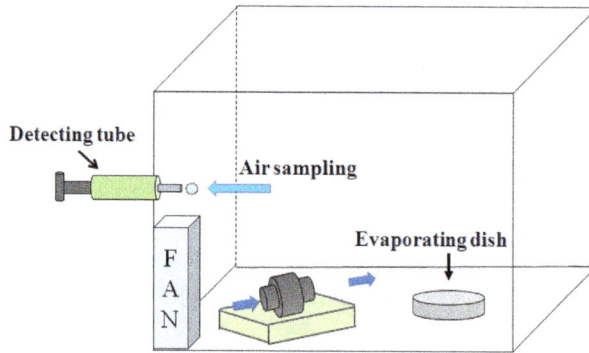

Figure 9. Experimental setup of ammonia removal in the experimental room (23.4 m³).

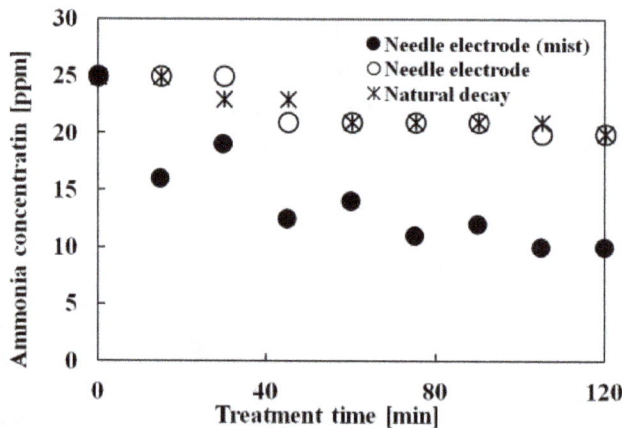

Figure 10. Ammonia concentration versus treatment time by the needle electrode.

During the 2 hours process, the ammonia concentration variation while energizing the needle electrode was the same as the concentration variation due to natural decay. This indicates that the needle electrode removed little ammonia. Note that the ozone concentration in the experimental room after two hours was 0.6 ppm. This relatively low ozone concentration resulted in no measureable reduction in the ammonia concentration. However, when the test was repeated with a fine water mist introduced into the needle electrode, ammonia removal reached 50%.

Well known gas phase reactions of ammonia with OH radicals and nitrogen oxides generate ammonium nitrate [18, 19].

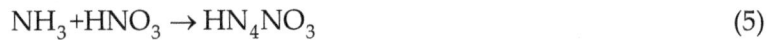

$$NO+OH+N_2 \rightarrow O+O_2+N_2 \tag{2}$$

$$NO_2+OH+N_2 \rightarrow OH+_3+N_2 \tag{3}$$

$$NH_3+HNO_2 \rightarrow HN_4NO_2 \tag{4}$$

$$NH_3+HNO_3 \rightarrow HN_4NO_3 \tag{5}$$

These reactions suggest why introducing a fine water mist into the needle electrode greatly increased rate of ammonia removal.

3.3. Sterilization of *E. Coli*

E. coli may be sterilized by the various active species formed by a microplasma discharge such as ozone, ions, and radicals [20]. Sterilization of *E. coli* (Migula 1895) was carried out using microplasma electrodes in the experimental room shown in Fig. 1. *E. coli* deposited on stamp media agar "Tricolor (El Mex, Inc.)" were placed at the seven positions shown in Fig. 11 in the experimental room.

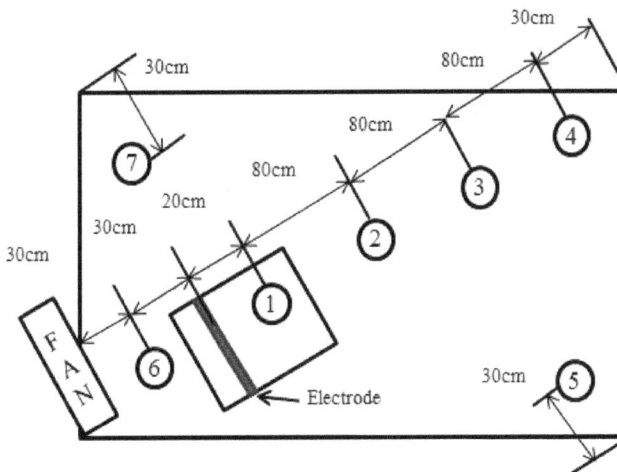

Figure 11. Position of the agar media placed in the experimental room.

The target bacteria were placed and treated with the microplasma for 2 hours. An 8th control medium with target bacteria was kept outside of the experimental room. Within the room, the microplasma electrodes were driven by a high-frequency alternating voltage of 27 kHz at voltages of 0.8, 0.9 and 1.0 kVp-p by an inverter and a neon transformer. After treatment, the 7 agar media and the 8th control medium were cultured for 24 hours in an incubator set to 37 °C. The sterilizing rate was obtained based on the control colonies number compared with the colonies number placed in the each position 1 to 7.

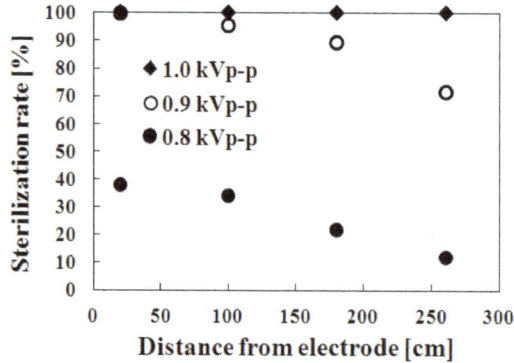

Figure 12. Sterilization rate for various discharge voltage at each position in the experimental room.

The sterilization rate increased with increasing voltage and decreased with distance from the electrode with an applied voltage of 0.8 and 0.9 kVp-p. There was no decrease of the sterilization rate with distance from the electrodes when applied voltage was 1.0 kVp-p. Increasing the voltage from 0.8 to 0.9 kVp-p resulted in a considerable increase in the sterilization rate. In the large experimental room, the effect of radicals is likely small since life time of radicals is short compared with the transit time of air moving from the reactor to the agar media test location. Consequently, radicals could not reach the target to sterilize [21, 22]. Figure 13 also shows the plasma treated *E. coli* for various distances from the electrode at applied voltage of 0.9 kVp-p.

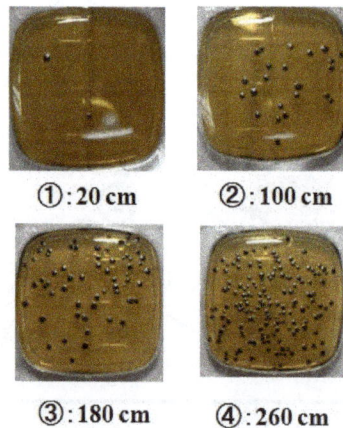

① : 20 cm   ② : 100 cm

③ : 180 cm   ④ : 260 cm

Figure 13. Plasma treated *E. coli* samples (Applied voltage 0.9 kVp-p).

We also measured the ozone and ions concentration at each position in Fig. 11 in the experimental room. Figure 14 shows the ozone and ions concentration of each position from the electrode, respectively. While the ozone concentration was very high at the outlet of the microplasma electrode, it decreased within a very short distance to a relatively uniform level that varied with voltage. On the contrary, ion concentrations varied strongly with position. This variation of ion concentrations with position was considerably different than the small chamber results. Ion concentrations depended on the distance from the microplasma electrode because ions have various reaction rate, and high reaction rate ions decayed with the distance by reacting with neutral molecules [23, 24].

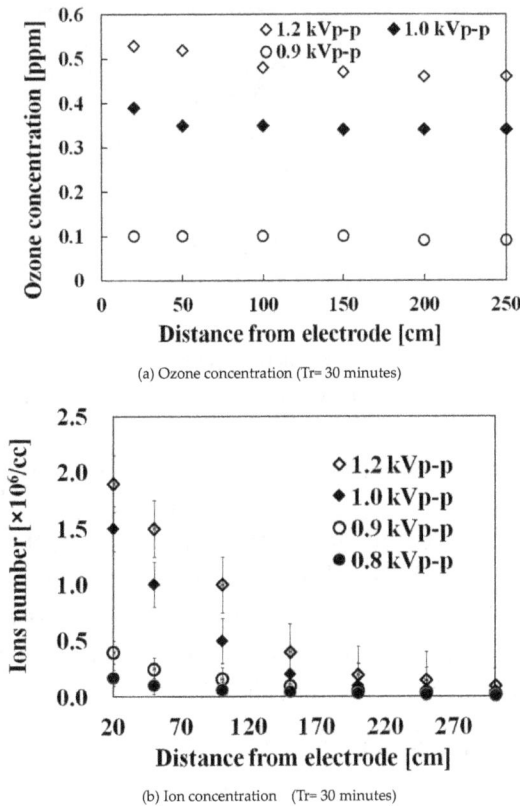

(a) Ozone concentration (Tr= 30 minutes)

(b) Ion concentration   (Tr= 30 minutes)

Figure 14. Ozone and ion concentration for various positions in the experimental room.

Even low concentrations of ozone in actual rooms could arise health issues [25, 26], so we investigated both ozone and ion concentration in the large experimental room. The variations of both ozone and ion concentrations are shown in Fig. 15. Ozone concentration increased with the process time, and exceeded the EPA regulatory [27] when the applied voltage was 0.9 kVp-p or higher. The concentrations of ions stabilized as shown in Fig. 15 during the process time (2 hours) at a level that depended on the applied voltage level.

The concentrations of ozone and ions shown in figures 14 and 15 suggest that the sterilization rates shown in Fig. 12 were related to the ozone concentration. However, relatively low

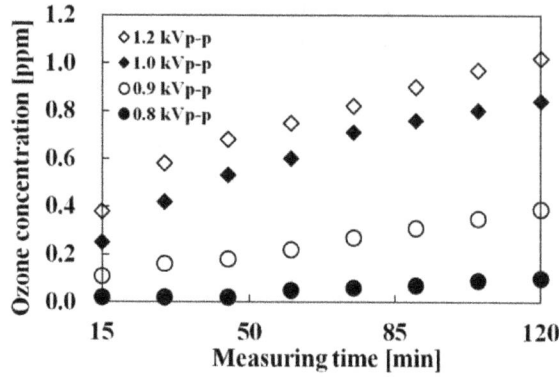

(a) Ozone concentration change (Position; 100cm away from electrodes)

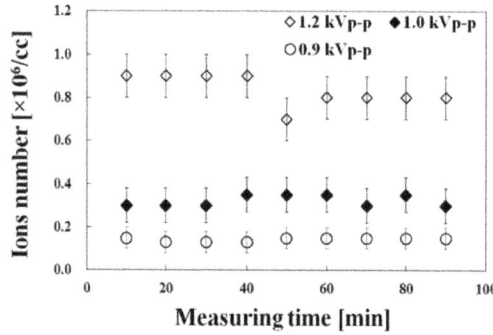

(b) Ions concentration change (Position; 100cm away from electrodes)

Figure 15. Variations in ozone and ion concentrations with microplasma treatment time in the experimental room.

concentrations of ozone are generally ineffective in sterilizing bacteria [28]. Similarly, the relatively low concentrations of ions diluted in the experimental room are also generally ineffective in sterilizing bacteria. Note that ions are only weakly reactive compared to highly reactive radicals. Consequently, there should be some reactions with reactive radicals that usually have short life time, such as OH radicals [21].

The ESR analysis of the agar medium in Fig. 16 with spin trap agent (5-dimethyl-1-pyrroline-N-oxide (DMPO)) indicate that the sterilization process may be the OH regeneration process shown in Fig. 17. Low concentration of ozone that reach the agar surface react as in equations (6) – (9) to generate OH radicals that sterilize the bacteria [29, 30].

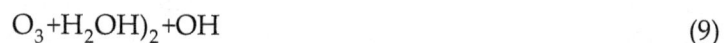

$$e + O_2 + OyO(^3P) + O(^3P,^1D) \tag{6}$$

$$O(^3P) + O_2 + M \ O)r_3 + M \tag{7}$$

$$O(^1D) + H_2O \rightarrow 2\,OH \tag{8}$$

$$O_3 + H_2OH)_2 + OH \tag{9}$$

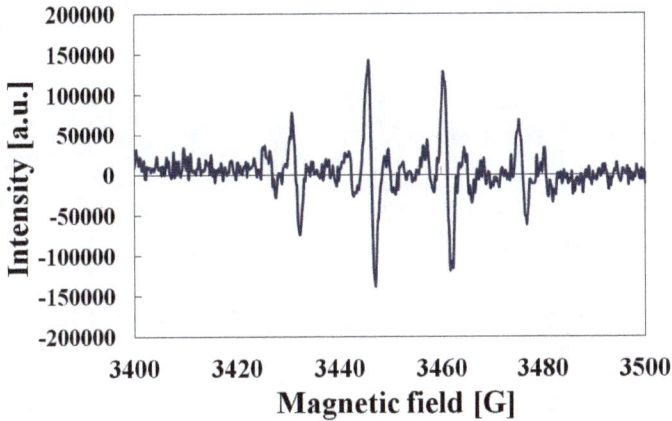

Figure 16. ESR analysis of the medium by plasma exposure.

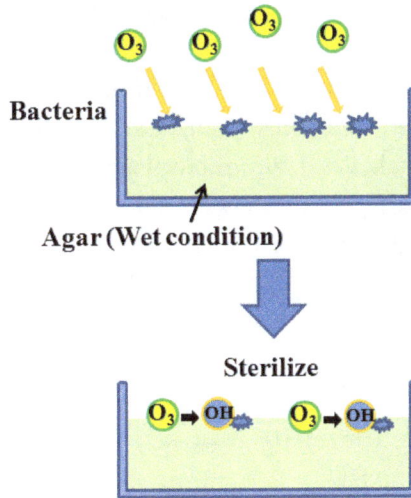

Figure 17. OH generation and sterilization process on the surface of the agar medium.

3.4. Sterilization of *E. coli* with ethanol and microplasma

In the large experimental room shown in Fig. 1, *E. coli* was sterilized by O_3 generated by a microplasma rather than by OH radicals whose short lifetime made them ineffective as a sterilization agent. We have enhanced the sterilization effect of bacteria in the small space experimentally as shown in Fig. 18.

A circular microplasma electrode (Diameter 50 mm) was installed in the experimental chamber having an inner diameter of 65 mm. *E. coli* was cultivated on the agar medium in the Petri dish. Air from the cylinder flowed at 10 L/min through a chamber containing liquid ethanol (70 vol %) that evaporated to generate 1.3% gas phase ethanol. The distance between the microplasma electrode and the Petri dish shown in Fig. 18 was 23 mm.

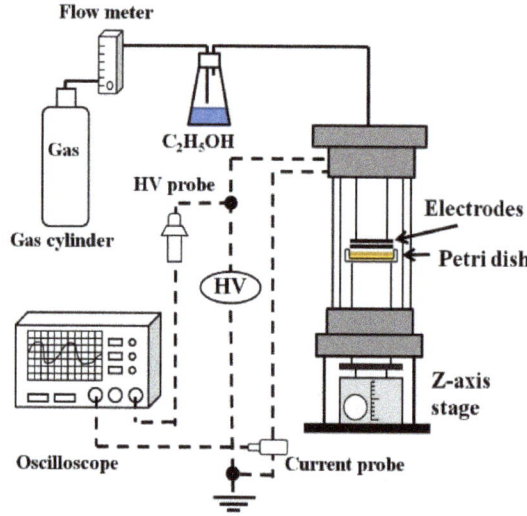

Figure 18. Experimental setup for sterilization of *E. coli* by microplasma with addition of ethanol.

With 1.3% gas phase ethanol, the microplasma sterilization process is greatly improved. Without microplasma treatment and only using ethanol, the decrease of bacteria colonies shown in Fig. 19 was minimal. With air microplasma treatment having no ethanol, the sterilization rate reached to 2digits (about 99%) as previously reported [6]. However, the microplasma treatment with 1.3% ethanol as an additive achieved a sterilization rate to 6digits with a 60 seconds treatment time.

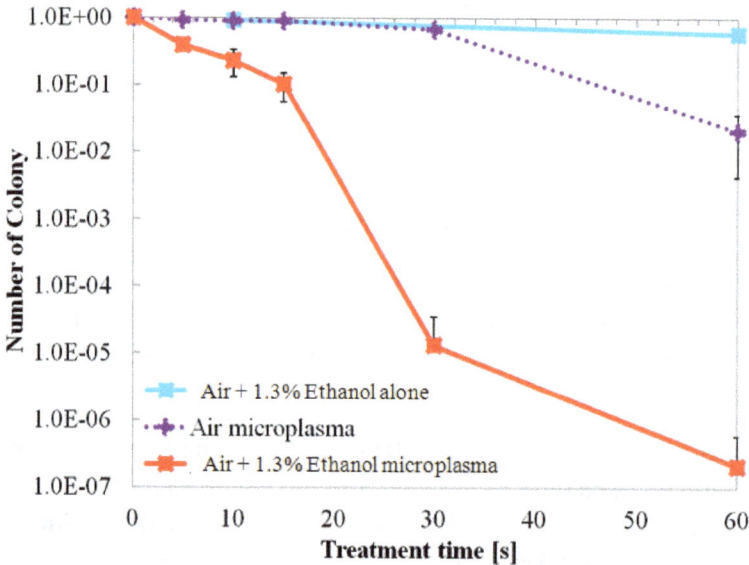

Figure 19. Sterilization of *E. coli* by microplasma and addition of ethanol.

For sterilizing E. coli in a large room such as the experimental room shown in Fig. 1, the life time of highly reactive radicals is too short for effective performance. Thus, experimental work was done to extend the distance between the microplasma electrode and the Petri dish in the small experimental chamber shown in Fig. 18. We found as shown in Fig. 20 that the sterilization performance of the microplasma with ethanol was effective at more than 20 cm away from the Petri dish. Relatively long life species must be generated by the microplasma with ethanol.

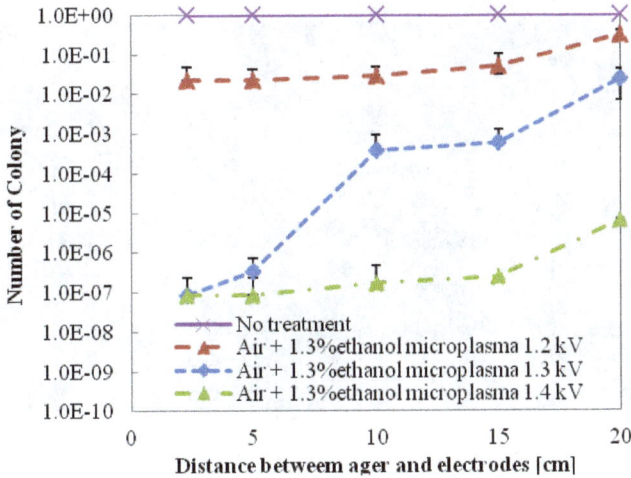

Figure 20. Sterilization rate dependencies with the distance between microplasma electrode and Petri dish.

Byproduct analysis of the process gas was carried out using FT-IR to identify the relatively long life species in the sterilization process. We found the undesirable byproducts shown in Fig. 21 with the ethanol additive process such as CO, and EOG (ethylene oxide gas). EOG is undesirable because it is well known to cause cancer. However, it should be noted that this process would be suitable for applications having a controlled room without any human [31].

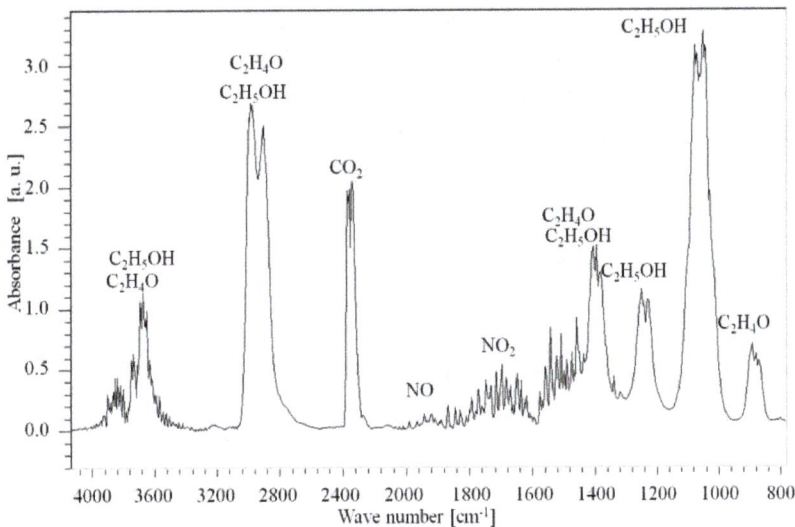

Figure 21. Byproduct analysis by FTIR for the microplasma treatment with addition of ethanol.

The shape of E. coli in the SEM images shown in Fig. 22 was completely changed by the microplasma process with ethanol addition. Note that this microplasma process is a remote plasma process. The electric fields and forces acting within the microplasma discharges are not acting on sample. Only various active species or chemical substances generated by the microplasma play a role in sterilizing the E. coli.

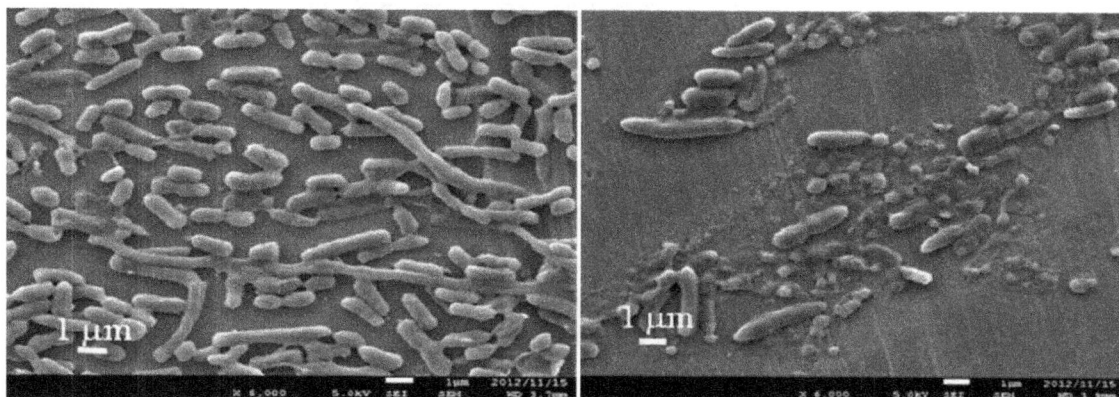

(a) Before treatment   (b) after 60 s of microplasma exposure

Figure 22. SEM images of microplasma treatment with ethanol addition (Vp=1.4 kV, treatment time; 60 s., ethanol concentration 1.3%, x 6, 000).

4. Conclusions

Since indoor air quality (IAQ) is so important for our health, indoor air purifiers are commercially available in Japan and in countries around the world. We have investigated the effect of atmospheric plasma including microplasma based on DBD technology both in a small experimental chamber and in a large experimental chamber that is the size of a room. Our results show that a low concentration of ozone played a role in the sterilization process and in decomposing harmful chemical substances such as ammonia. Conducting mesh and electrostatic filters also played a role in collecting particulate material (PM), which are fine particles.

Plasma treatment technologies are relatively new and their performance remains uncertain. Atmospheric plasma can generate ozone various kinds of radicals, ions, and other active species. We propose various plasma treatments for applications to improve the public health. Commercially available plasma devices based on the corona discharge technology and electrostatic precipitation generate low concentrations of ozone. Our results show that ozone at low concentrations is a surprising effective sterilization agent. However, the academic sector has an important responsibility to public health to reveal that plasma processes can also generate harmful byproducts. Our results with investigating plasma technologies to improve the indoor air quality show great promise. It is our pleasure to understand the usefulness of atmospheric microplasmas and their applications

Acknowledgements

The author thanks to the Dr. Marius Blajan, and Mr. Yusuke Kurokawa of Shizuoka University for fruitful discussions. The author also thanks to Mr. Masataka Sano of the Ceraft Co. Ltd. for their technical support.

Author details

Kazuo Shimizu*

Address all correspondence to: shimizu@cjr.shizuoka.ac.jp

Shizuoka University, Japan

References

[1] Vanda Villanyi, Air Pollution, Intech; 2010.

[2] Michael Theophanides, Jane Anastassopoulou and Theophile Theophanides. Air Polluted Environment and Health Effects. Prof. José Orosa (Ed.) Indoor and Outdoor Air Pollution. ISBN: 978-953-307-310-1, InTech; 2011

[3] Silvia Vilcekova. Indoor Nitrogen Oxides. Dr. Farhad Nejadkoorki (Ed.) Advanced Air Pollution. ISBN: 978-953-307-511-2, InTech; 2011.

[4] Ricardo Araujo and João P. Cabral. Fungal Air Quality in Medical Protected Environments. Ashok Kumar (Ed.) Air Quality. ISBN: 978-953-307-131-2, InTech; 2010.

[5] Risto Kostiainen. Volatile Organic Compounds in the Indoor Air of Normal and Sick Houses. Ahnospheric Enuironment 1995; 29 (6) 693702.

[6] Kazuo Shimizu. Indoor Air Control by Microplasma. Dr. Farhad Nejadkoorki(Ed.) Advanced Air Pollution, ISBN: 978-953-307-511-2, InTech; 2011.

[7] Heidi Ormstad. Suspended particulate matter in indoor air: adjuvants and allergen carriers. Toxicology 2000; 152 53–68.

[8] Markus Sillanpää, Michael D. Geller, Harish C. Phuleria, Constantinos Sioutas. High collection efficiency electrostatic precipitator for in vitro cell exposure to concentrated ambient particulate matter (PM). Aerosol Science 2008; 39 335 – 347.

[9] Shoji Koide, Akira Nakagawa, Katsuhiko Omoe, Koichi Takaki, Toshitaka Uchino. Physical and microbial collection efficiencies of an electrostatic precipitator for abat-

ing airborne particulates in postharvest agricultural processing. Journal of Electrostatics 2013; 71 734-738.

[10] M. Blajan, and K. Shimizu. Phenomena of Microdischarges in Microplasma. IEEE Trans. on PS 2012; 40 (6) 1730-1732.

[11] M. Blajan, and K. Shimizu. Temporal evolution of dielectric barrier discharge microplasma. Appl. Phys. Lett 2012; 101 104101.

[12] L. F. Gaunt, J. F. Hughes, N. M Harrison. Removal of domestic airborne dust particles by naturallycharged liquid sprays. J. Electrostat 2003; 58 159-169.

[13] G. D. Conanan, F. C. Lai. Performance Enhancement of Two- Stage Corona Wind Generator in a Circular Pipe. Proc. 2012 Electrostatics Joint Conference, Cambridge, K3, June 12-14, 2012.

[14] K. Shimizu, Y. Kurokawa, M. Blajan. Basic Study on Indoor air Quality improvement by Atmospheric Plasma. IEEE IAS Annual Meeting 2014; 2014-EPC-0308, 2014.

[15] K. Shimizu, T. Kuwabara, and M. Blajan. Study on Decomposition of Indoor Air Contaminants by Pulsed Atmospheric Microplasma. Sensors 2012. 12. 14525-14536.

[16] Jan Mertens, C. Anderlohr, P. Rogiers, L. Brachert, P. Khakhariac, E. Goetheer, K. Schaber. A wet electrostatic precipitator (WESP) as countermeasure to mist formation in amine based carbon capture. International Journal of Greenhouse Gas Control 2014; 31 175–181.

[17] Yih-Juh Shiau, Chien-Young Chu. Comparative effects of ultrasonic transducers on medium chemical content in a nutrient mist plant bioreactor. Scientia Horticulturae 2010; 123 514–520.

[18] L. Komunjer, C. Affolter. Absorption–evaporation kinetics of water vapour on highly hygroscopic powder: Case of ammonium nitrate. Powder Technology 2005; 157 67 – 71.

[19] L. Xia, L. Huang, X. Shu, R. Zhang, W. Dong. Removal of ammonia from gas streams with dielectric barrier discharge plasmas. Journal of Hazardous Materials 2008; 152 (1) 113-119.

[20] K. Shimizu, M. Blajan, and S. Tatematsu. Basic Study of Remote Disinfection and Sterilization Effect by Using Atmospheric Microplasma. IEEE Trans. on Ind. Appl. 2012; 48 (4) 1182-1188.

[21] Ryo Ono, Tetsuji Oda. Measurement of hydroxyl radicals in pulsed corona discharge. Journal of Electrostatics 2002; 55 333–342.

[22] S. Kanazawa, H. Tanaka, A. Kajiwara, T. Ohkubo, Y. Nomoto, M. Kocik, J. Mizeraczyk, Jen-Shih Chang. LIF imaging of OH radicals in DC positive streamer coronas. Thin Solid Films 2007; 515 4266–4271.

[23] K. Nagato, Y. Matui, T. Miyata, T. Yamauchi. An analysis of the evolution of negative ions produced by a corona ionizer in air. Int. J. Mass Spectrom 2006; 248 (3) 142-147.

[24] K. Sekimoto. Influence of needle voltage on the formation of negative core ions using atmospheric pressure corona discharge in air. Int. J. Mass Spectrom 2007; 261 (1) 38-44.

[25] Charles J. Weschler. New Directions: Ozone-initiated reaction products indoors may be more harmful than ozone itself. Atmospheric Environment 2004; 38 5715–5716.

[26] William W. Nazaroff, Charles J. Weschler. Cleaning products and air fresheners: exposure to primary and secondary air pollutants. Atmospheric Environment 2004; 38 2841–2865.

[27] U.S. Environmental Protection Agency, Office of Air and Radiation, Office of Air Quality Planning and Standards. Regulatory Impact Analysis -Final National Ambient Air Quality Standard for Ozone-. 2011.

[28] F. M. Gabler, J. L. Smilanick, M. F. Mansour, H. Karaca. Influence of fumigation with high concentrations of ozone gas on postharvest gray mold and fungicide residues on table grapes. Postharvest Biology and Technology 2010; 55 85-90.

[29] B. Eliasson, M Hirth and U. Kogelschatz. Ozone synthesis from oxygen in dielectric barrier discharges. J. Phys. D: Appl. Phys. 1987; 20 (11) 1421-1437.

[30] H. S. Ahn, N. Hayashi, S. Ihara and C. Yamabe. Ozone Generation Characteristics by Superimposed Discharge in Oxygen-Fed Ozonizer. J. Appl. Phys 2003; 6578- 6583.

[31] P.A. Schulte, M. Boeniger, J.T. Walker, S.E. Schober, M.A. Pereira, D.K. Gulati, J.P. Wojciechowski, A. Garza, R. Froelich, G. Strauss, W.E. Halperin, R. Herrick and J. Griffith. Biologic markers in hospital workers exposed to low levels of ethylene oxide. Mutation Research/Genetic Toxicology 1992; 278 237-251.

8

Treatment of Post-Consumer Vegetable Oils for Biodiesel Production

Elaine Patrícia Araújo, Divânia Ferreira da Silva,
Shirley Nobrega Cavalcanti,
Márbara Vilar de Araujo Almeida,
Edcleide Maria Araújo and Marcus Vinicius Lia Fook

Additional information is available at the end of the chapter

1. Introduction

The current energy model based on petroleum shows signs of exhaustion, which is aggravating, as besides energy source petroleum is used extensively for the production of plastics, clothing, fertilizers and medicine, moving a true "Petroleum Civilization" [1]. Ally the question of exhaustion of petroleum reserves and its derivatives and the search for renewable energy sources, is also highlighted the issue of waste, which daily becomes one of major problems for humanity. Worldwide, approximately 60 million tons of edible vegetable oils - which, in most cases, are used for frying various types of food - are produced, according to data from the United States Department of Agriculture Food, published in 2000. A significant number of these oils are eliminated directly into the environment, harming these aquatic and terrestrial environments [2].

Almost all energy consumed in the world comes from non-renewable sources of fossil fuels, which cause great environmental impact. Alternative fuels for diesel engines are becoming increasingly significant due to the decrease of petroleum reserves and thus, increasing it's price, that reaches levels high enough to prevent it's use. Also the environmental impact caused by emissions of gases generated from burning of fossil fuels have been reason for research on alternative energy sources [3,4].

Due to emission of toxic gases by discharges from diesel vehicles, hundreds of researches warn that different pollutants emanating from the exhausts lie in the main causes of degradation of air quality in large urban centers.

The recycling of post-processed oil is minimal and has restricted applications, one being the use in the detergent industry and most recently, as biodiesel. This can be defined as the mono-alkyl ester derived from long chain fatty acids, from renewable sources such as vegetable oils or animal fats obtained by transesterification process, use of which is associated with replacement of fossil fuels in engines compression ignition [5]. It can also be defined as a biodegradable fuel derived from renewable resources obtained from the reaction of vegetable oil and animal fats which, stimulated by a catalyst, react chemically with methanol or ethanol. This can be done with any fresh or post-consumer vegetable oil, waste or sludge. Several studies have shown that the obtaining of methyl and ethyl esters from soybean oil, canola, sunflower, palm, castor, and also post-consumer frying oil, is recommended, since it has lower incomplete combustion of hydrocarbons and lower emissions of carbon monoxide, particulate matter, nitrogen oxides and soot [3, 6].

The biodiesel obtained from post-consumer frying oil, according to studies, decreased smoke, demonstrating that has effective benefit in reusing this oil for biofuel production, featuring a more suitable destination to this agro-industrial waste that, in Brazil, is commonly discarded and/or partially reused, but often in inadequate ways [7].

Vegetable oils have many advantages as alternative fuels when compared to diesel: they are natural liquids, renewable, with high energy value, low sulfur content, low aromatic content and biodegradable. However, despite the use of these oils being favorable from the point of view of energy, its direct use in diesel engines is very problematic. Studies performed with various vegetable oils showed that its direct combustion leads to a series of problems: carbonisation in the injection chamber, contamination of the lubricating oil, among others [8].

The emission of toxic gases by motor vehicles is a major source of air pollution. In cities, these vehicles are responsible for the emission of harmful gases such as carbon monoxide (CO), carbon dioxide (CO_2), nitrogen oxides (NO_x), sulfur dioxide (SO_2), hydrocarbons (HC), lead, smoke and particulates. Studies have been conducted in order to quantify and estimate the use of various energy sources on the increase of CO_2. The main sources of energy considered more polluting in terms of CO_2 emissions are: Liquefied Petroleum Gas (LPG), natural gas, fuel oil and diesel oil [9].

In general, air pollution affects health, generating both acute effects such as eye irritation and coughing, which are temporary and reversible, and chronic effects, which are permanent and cumulative with demonstrations in the long run, of causing severe respiratory diseases. There may also be structural corrosion and degradation of buildings and work of art. In heavily polluted cities, these disturbances are exacerbated in winter with temperature inversion, when a layer of cold air forms a bell high in the atmosphere, trapping hot air and preventing dispersion of pollutants. Compared to different sources of emissions, the diesel has the highest emission of toxic gases, contributing to the rise of the various environmental scenarios, social and economic [10]. Due to this problem, various studies are being conducted with post-consumption vegetable oils for biodiesel production, therefore, is an alternative renewable fuel that releases less harmful gas emissions compared to conventional fossil fuel (diesel). The most common method of making biodiesel is by transesterification reaction of vegetable oils or animal fats, with a short chain alcohol [11].

Biodiesel in its pure form (B100) can allow the net emission of carbon dioxide (CO_2) - the main from the Greenhouse Gas (GHG) emissions be reduced by 80%. This has a positive impact on the environment because it decreases air pollution in large urban centers (the B100 blend provides a 90% reduction of smoke and eliminate the sulfur oxide, responsible for acid rain), thus improving the quality of life and reduced spending in the health system population [12].

The recycling of post-consume discarded vegetable oils contributes to reduce the uncontrolled and harmful environmental disposal, and may have competitive price on fossil fuels. However, the use of this oil in biodiesel production requires treatment prior to transesterification reaction, which comprises the removal of contaminants solid particles and the appropriateness of color and odor. For this reason, the main objective in this research was to use clays of the northeastern semi-arid region of Paraíba/Brazil to evaluate its potential in the treatment of post-consumer vegetable oils for biodiesel production.

1.1. Post-consumer vegetable oils

Among renewable energy resources, the use of biomass, in its different forms (solid, liquid and gas), was intensively researched in recent years as an alternative to minimize adverse environmental impact and the uncertainty in future supply of fossil fuels. Despite the possible environmental benefits in the use of vegetable oils as a substitute for diesel, barriers economically and ethically motivated the search for alternative raw materials for biofuel production [13, 14].

Among the alternatives studied, the reuse of Waste Vegetable Oils (WVO) and fats of processes of frying various foods has been shown to be attractive because the advantage in that the vegetable oil as fuel after its use in the food chain, thus resulting in a second use, or even an alternative destination to a residue of food production destination [15]. Among the several aspects that motivated the study of vegetable oils as fuel potential are:

- It's (liquid) physical condition and its high specific energy content (MJ/kg fuel) when compared with other fuels derived from biomass;

- The fact that they are produced from different oilseeds (soybean, rapeseed, palm, etc.) under different climatic conditions;

- High energy productivity of some oilseeds (above 150 GJ/ha for palm oil);

- The possibility of using oil and its derivatives in high efficiency energy conversion engines, such as diesel engines.

The oils used in frying have important nutritional aspects, involving the transport of fat soluble vitamins, supply of essential fatty acids of the $\omega3$ and $\omega6$ series, precursors of eicosanoids, the energetic power and present a wide acceptance by the various social groups [16].

Studies conducted in the city of Valencia/Spain, concludes that it is attractive, the environmental point of view, obtaining biofuel from WVO. A selective collection system, established by the city council, supported the project to produce biofuel to supply 480 city buses, with a demand of approximately 42,000 liters/day. The ultimate goal of the project was the elimination, large scale, the WVO plumbing the sanitary sewer system of the city, about 10,000 t/a [13].

In Brazil, it is common the use of soybean oil (nationally) and rice oil (in the south) to processes of frying food in shops. Soybean oil contains 15% of saturated fatty acids, 22% of oleic acid, 54% of linoleic acid and 7.5% linolenic acid. The rice oil contains about 20% of saturated fatty acids, 42% of oleic acid, 36% of linoleic acid and 1.8% of linolenic acid. Soybean oil, by presenting a lower composition of saturated fatty acids and higher in polyunsaturated fatty acids, is more susceptible to degradative processes [17].

The physical changes which occur in the oil or fat during the frying process include dimming, increase in viscosity, decrease in smoke point and foaming. Chemical changes can occur by three different types of reactions: the oils and fats can hydrolyze to form free fatty acids, monoacylglycerol and diacylglycerol; can oxidize to form peroxides, hydroperoxides, conjugated dienes, epoxides, ketones and hydroxides; and may decompose into small fragments or remains in the triacylglycerol and to associate, leading to dimeric and polymeric triglycerides [16,18].

During the process of frying oils and fats are exposed to the action of three agents that contribute to compromise their quality and modify its structure: the moisture from foods, which is the cause of hydrolytic alteration; oxygen, which in contact with the oil, for prolonged periods, causes oxidative modification and the high temperature in the operation, 180°C, causing thermal alteration [19].

The usage time of the oil varies from one establishment to another, mainly due to the lack of legislation to determine the exchange of post-consumer oil [7]. There is no single method by which it is possible to detect all situations involving the deterioration of oils in the frying process. The determination of the optimal point for disposal has significant economic impact resulting in a higher cost when oil is discarded before its effective degradation, and loss of quality of food, when discarded later. Some indicators used by restaurants and cafeterias, to determine the point of discharge of oil or fat are: color change, formation of smoke and foam during the frying process and changes in aroma and taste [20].

In frying temperature (170 to 180°C) occurs in reactions with air, water and food components. The oil and the vegetable fat used in frying process by immersion represent a major risk of environmental pollution since most commercial establishments (pubs, restaurants, coffee shops, etc.) and residential discard the residual oil into the sewer system difficult to treat these. However, this material can be used as raw material for biodiesel production [21, 10].

The transformation of the used cooking oil into biodiesel brings significant environmental improvements. Initially, the byproduct that would be discarded in the environment receives a new use, no longer willing improperly. Thus, reducing the consumption of fossil fuels (diesel oil) occurs, in addition to encouraging the use of renewable fuels.

For the manufacture of biodiesel, it is necessary to invest in the industry of purification and transformation. Biodiesel is biodegradable fuel derived from renewable sources (vegetable oil or animal fat) which can be obtained by different processes, such as cracking, esterification and transesterification [22].

The lipids are oils and fats insoluble in water, animal or vegetable source, and consist of triglycerides or triglyceride esters formed from glycerol and fatty acids. The present fatty acids are generally saturated carboxylic acids with 4 to 24 carbon units in the chain and unsaturated carboxylic acids with 10 to 30 carbons and 1 to 6 double bonds in the chain [23].

The vegetable oils are natural products consist of a mixture of esters of glycerol derivatives, which contain fatty acid chains from 8 to 24 carbon atoms having different degrees of unsaturation (Figure 1). Different species show variations in oil molar ratio between the different fatty acids present in the structure [7].

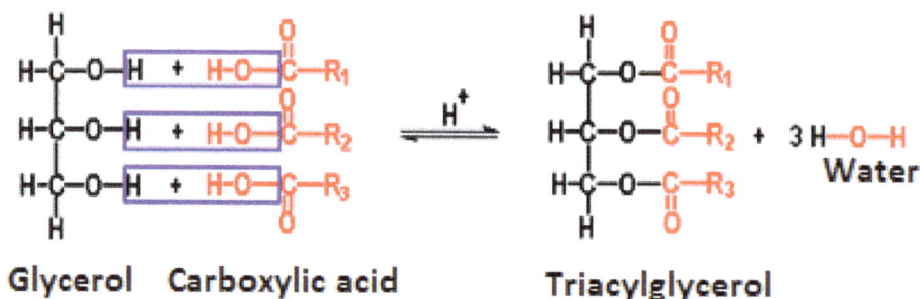

Source: [24].

Figure 1. Formating of Trigliceride: a molecul of glicerol and a molecul of fatty acid.

Firestone et al. [25] comment that in some countries such as Belgium, France, Germany, Switzerland, Netherlands, United States and Chile, there are rules on the conditions under which a vegetable oil used for frying should be discarded. But in Brazil, as in many other countries there are no laws and regulations establishing limits to the changes in these oils. An estimated damaged by oil frying process must be discarded when their content of polar compounds meet above 25%. Another aspect that must be considered is the percentage of free fatty acids, for which the laws set, limits around 1 and 2.5%.

One of the main causes of the degradation of oils and fats is rancidity, which is associated with the formation of organoleptically, creates unacceptable product due to occurrence of foreign odors and flavors, and the loss of product color, and inactivation of vitamins polymerization [26]. The rancidity can be classified as:

• Hydrolytic rancidity - occurs in the presence of moisture due to the action of lipases that catalyze the hydrolysis enzymes, releasing fatty acids, and

• Oxidative rancidity or oxidation occurs due to non-enzymatic action of lipoxygenase enzymes or by action, such as autoxidation and photooxidation.

According [17] consumption of fried foods and frozen pre-fried, induces higher intake of oil through the frying process. During these processes, there are several forms of lipid deterioration that compromises the quality of the oil, they are:

- Hydrolysis involves the cleavage of the glyceride ester with formation of free fatty acids, mono glycerides, diglycerides and glycerol. It is a reaction that occurs due to the presence of water at high temperatures, which can result in products with high volatility and high chemical reactivity;

- Consisting of degradative oxidation process in which atmospheric oxygen dissolved in the oil or reacts with unsaturated fatty acids, producing sensory unacceptable products with unpleasant smells and flavors for human consumption (Figure 2);

- Polymerization that occurs when two or more fatty acid molecules combine as a consequence of changes in the oxidation process and high temperatures.

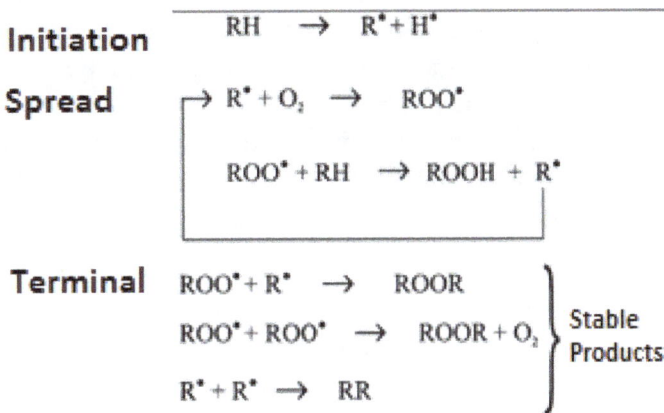

Initiation $RH \rightarrow R^{\bullet} + H^{\bullet}$

Spread $R^{\bullet} + O_2 \rightarrow ROO^{\bullet}$

$ROO^{\bullet} + RH \rightarrow ROOH + R^{\bullet}$

Terminal $ROO^{\bullet} + R^{\bullet} \rightarrow ROOR$

$ROO^{\bullet} + ROO^{\bullet} \rightarrow ROOR + O_2$ Stable Products

$R^{\bullet} + R^{\bullet} \rightarrow RR$

Source: [27].

Figure 2. General scheme of the lipid oxidation mechanism.

Several studies with oil heated for long periods at high temperatures showed that the resulting product contains more than 50% of polar compounds which are degradation products of triglycerides (polymers, dimers, oxidized fatty acids, diglycerides and free fatty acids). These oils with high contents of polar compounds can cause severe irritation of the gastrointestinal tract, diarrhea, reduction in growth, and in some cases death of laboratory animals [28]

When oils are used at high temperatures or are reused, they release a toxic substance, acrolein, which interferes with the functioning of the digestive and respiratory system, mucous membranes and skin, and can even cause cancer [29].

The resulting polymers increase the viscosity of the oil. The frying process characteristics such as browning develops, an increase in viscosity, decrease in smoke point and foaming affecting the quality of the oil [16].

Brazil does not have any regulation that legally defines the monitoring of disposal for oil and frying fats. There are regulations governing the suitability of oil for consumption in Brazil, the NTA 50, citing some physicochemical items to control the suitability of this oil: iodine value, peroxide value and acid value, however not refer to oils and cooking fats [16].

1.2. Clays

According to Santos [30], clays are natural earth materials which exhibit fine-grained (typically with a diameter of less than 2 μm particles) and are formed by chemically hydrated silicates of aluminum, iron and magnesium. These are composed of small crystalline particles of a limited number of minerals, clay minerals. In addition to these clay minerals, clays may also contain organic matter, soluble salts, particles of quartz, pyrite, calcite, and other residual amorphous mineral reserves. The main factors that control the properties of clays are the mineralogical and chemical composition of clay minerals of non-clay minerals and their particle size distributions; electrolyte content of exchangeable cations and soluble salts; nature and content of organic components and textural characteristics of the clay.

Brazil has industries that utilize different types of clays for several purposes: fabrications red ceramic, white ceramic, refractory materials; the manufacturers of rubber and plastics used as the active and inert fillers; metallurgical industry uses clays as binders for molding sands for the casting of metals and for pelletizing iron ores; industries of edible oils and petroleum use them as bleaching agents of vegetable and mineral oils; can also be used as thixotropic agents in mud for drilling for oil drilling and water; There are special clays are used as catalysts in cracking of oil to produce gasoline and are used for special purposes being used as filler for soap and tissues, as pigments for paints, in the manufacture of pharmaceutical products [30].

Determining the result of the technological properties of these properties whose function is to complement function test results traditional characterization as: X-ray diffraction, X-ray fluorescence, particle size analysis. With these results together with the results of the technical properties (physical and mechanical properties) can indicate the proper use of a clay and establish accurate or necessary for better performance properties to which the clay is subjected [31].

The importance and diversity of use of clays is a result of its particular characteristics. This difference makes the clays of the most used materials, either on his great geological variety or offers a set of essential and indispensable factors in numerous industrial processes [31].

The bentonite is a layered clay mineral composed montmorillonite that is an aluminosilicate trifórmico the type crystalline structure appearing as a layer of alumina octahedrons between two layers of silica tetrahedra with adjacent margins primarily. Their composition is variable due to ease of isomorphic substitutions (may contain FeO, CaO, Na_2O and K_2O), which causes a negative charge density on the surface of the smectite clay and require cations to compensate for these loads, the exchangeable cations [30].

In Brazil, the terms bentonite clay materials are used to montmorillonite without any information about the geological origin or mineralogical composition. The chemical composition and method of the unit cell of the "theoretical" montmorillonite or end of the series is $(Al_{3,33}Mg_{0,67})\ Si_8O_{20}\ (OH)\ 4.M^{+}_{10.67}$, where M^{+1} is a monovalent cation. This formula shows that the unit cell has a negative electrical charge due to isomorphic substitution of Al^{3+} by Mg^{2+}. The cation M^+ which balances the negative charge is called exchangeable cation since it can be changed in a reversible way, by other cations. The content of exchangeable cation, expressed in milliequivalents of cation per 100g of clay is called CEC - cation exchange capacity. The cation M^+ interplanar occupies the space of the two layers 1 and may be anhydrous or hydrated. As the size of the dry cation and the number of layers of water molecules coordinated to the cation, it may have different values of basal interplanar distance [32].

According Centre for Mineral Technology – CMT [33], is the term given to a smectite group of minerals consisting of: montmorillonite, beidellite, nontronite, hectorite and saponite, in which each of these minerals form a similar structure, but each is chemically different. The most common mineral in the economic deposits of smectite is montmorillonite group. The calcic and sodic varieties, based on exchangeable cation, are the most abundant.

Amorim et al. [34] commented that according to geologists, the bentonite is formed by devitrification and chemical alteration of volcanic ash. For many years, scholars have used the origin of these clays as part of its definition, but in some countries as their deposits were not originated by volcanic action, other definition came to be used: bentonite clay is composed of any clay mineral montmorillonite, smectite group the and whose properties are established by this clay mineral.

The bentonite clays have distinct and peculiar to increase to several times its original volume when wetted with water and form thixotropic gels in aqueous media at low concentrations, interplanar spaces reaching up to 100 Å, high surface area and cation exchange capacity. These are characteristics that make the bentonite a wide range of applications in various technological sectors from the preparation of nanocomposites by the use as decolorizing agent [35].

Deposits of bentonite clays of Paraíba form the largest, and the most important deposit is located in Brazil. Their occurrences are located in the city of Boa Vista, and its deposits are mines Lages, Bravo, Jua and Canudos. In 2004, Paraíba State was the main producer of crude bentonite with 88% of national production, followed by São Paulo (7.3%), Rio de Janeiro (4.4%) and Paraná (0.2%). The production of bentonite in Brazil, which focuses on two products, activated bentonite clay and dry ground, grew by 14% [34, 36].

According to data released by the National Department of Mineral Production – NDMP [37], the state of Paraíba is currently the most significant source of bentonite clay, bentonite deposits it's being located mainly in the city of Boa Vista. Its reserves amount to about 70% of bentonite clays throughout Brazil.

The national reserves of bentonite represent about 3% of world reserves. Brazilian production is around 300 000 t/a which represents 3% of world consumption. The average price of bentonite is about $107/t, while the activated bentonite can reach $1,800/t. Also according to [38], the market for bentonite is very concentrated in the United States, the world's largest producer and has high investments made in this industry, which has provided diversification in its use and application [38].

Clay minerals of the smectite (montmorillonite) group are composed of two layers of a tetrahedral silicate with an octahedral core sheet joined together by common oxygen atoms to the leaves and in the space between the sheets are adsorbed water molecules and exchangeable cations, which may be Ca^{2+}, Mg^{2+} or Na^+ or both (Figure 3) [30].

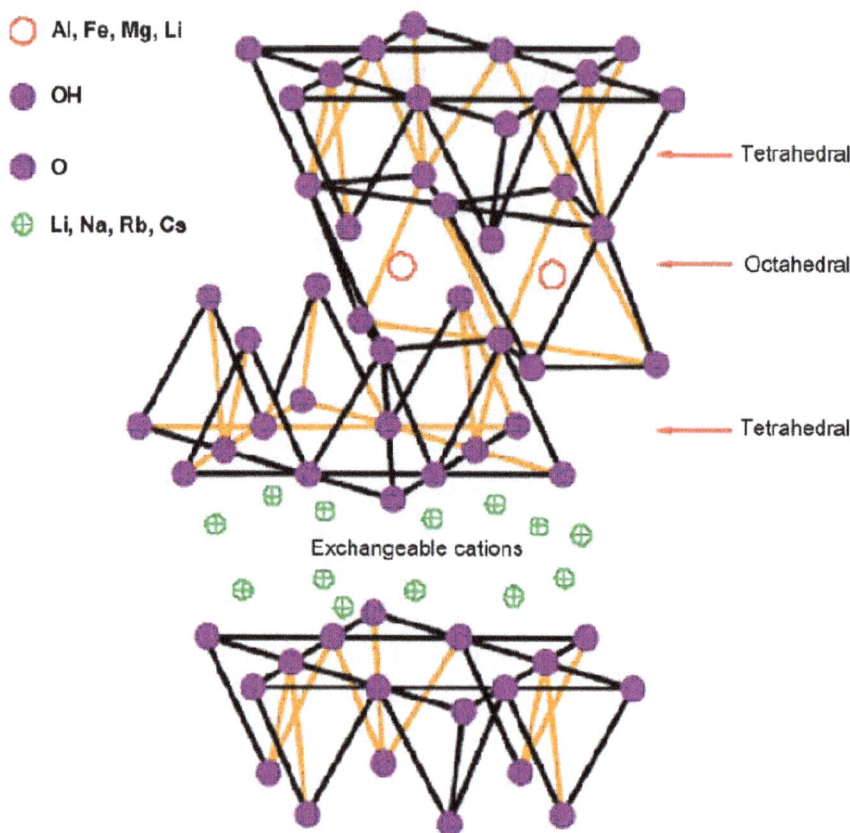

Source: [39, 40].

Figure 3. Crystal structure of the clay mineral montmorilonitico.

The bentonite clay is classified according to their exchangeable cations present in [39]:

- Homocationica: when there is a predominance of a type of exchangeable cation such as sodium or calcium, so called sodic or calcic bentonite.

- Polycationic: when there is no predominance of one type of exchangeable cation. Cations such as sodium, calcium, potassium and others can be present in similar concentrations.

Treatment with acid serves to Dissolve some impurity of bentonite; replace calcium and other cations intercalated by ions H_3O^+ hidroxônio and dissolve in the octahedral layers of two layers: one, some cations Mg^{2+}, Al, Fe^{3+} or Fe^{2+}. The acid treatment causes significant morphological changes in the crystal structure of montmorillonite during and after acid activation. The montmorillonites activated by acids, are commonly used for bleaching of edible oils and fats [40].

1.3. Clays for treatment of post-consumer vegetable oils

Clays have been used by mankind since antiquity for manufacturing ceramic objects, such as bricks and tiles, and more recently, in several technological applications. These are used as adsorbents in bleaching processes in the textile and food industry, in processes of soil reme-diation and landfill. The interest in its use has been gaining momentum mainly due to the search for materials that do not harm the environment when discarded, the abundance of world reserves and its low price. In the oil industry, the clays that are used for bleaching these oils are called "bleaching earth", "soil bleach", "clarifying clay" or "adsorbent clay" to indicate that clays in the natural state or after chemical or thermal activation, have the property of coloring materials present in adsorbing mineral oils, animal and vegetable [32].

The bentonite clays according to Santos [32] can be classified according to their adsorptive properties: montmoriloniticas bentonite-type clays, which are virtually inactive and inativa-veis ; montmoriloniticas inactive clays, but highly activatable by acid treatment; extremely active and activatable clays by acid treatment; active clays and whose activity is little affected by acid treatment; active clays whose activity is decreased by acid treatment.

According to the adsorptive and catalytic properties, the activated bentonite clays are used industrially as catalysts, adsorbents and catalyst supports. However, in terms of consumption, the most important use of this material and purification, bleaching and stabilization of vegetable oils. The adsorptive capacity of these materials increases with treatment with strong acid, typically sulfuric or hydrochloric acid are used. The presence of these acids modifies the structure of clays [41].

The adsorptive capacity of clay bleaching increases with the increase of the specific area. The bleaching earth adsorbs some better connection than others or even ceases to adsorb some. Polar or polarizable molecules are well adsorbed by bleaching earth. However, the adsorptive ability of the bleaching earth is reduced if the oil contains soaps or gums in excess to neutralize the acid sites of the same is true when there are many free fatty acids, which, as highly polar compounds occupy part of the surface of the clay mineral [26].

The power of bleaching clay may be due alone or in combination, the following factors: simple filtration, which corresponds to the retention of colored particles dispersed in oil in the capillaries clay; the selective adsorption of dissolved dyes and catalytic activity of the clay [32].

The time of bleaching oils suffer limitations due to the bleaching temperature. For this, we used 0.75% of smectite clays activated bleaching processes structured in three different temperature levels (82°C, 104°C and 138°C) and five levels of time (5 min., 10min., 15 min., 35min., and 55min.). It was observed that the red color of the oil fell to the lowest level when the highest temperature is used. At this temperature, however, the color began to darken oil from the time of bleaching, coming, at the end of the bleaching, become darker than the other two processes [26].

For an acid activated bentonite clay may be used as a decolorizing agent is necessary to have the following requirements: the pH is between 6.0 and 7.5; porosity between 60 and 70%; no catalytic activity in the case of edible oils and fats to prevent the generation of undesirable odors and tastes after bleaching; low oil retention in filtration and good filterability [40].

2. Materials and methods

The calcic clays used for the treatment of post-consumer vegetable oils were the bentonite clay (Figure 4) trade name Tonsil and Aporofo with a particle size of 200 mesh (0.074mm) mesh provided and identified by the company BENTONISA - The Bentonite Nordeste S/A, located in João Pessoa-Paraíba.

Source: Research data.

Figure 4. Calcium bentonite clay used for the treatment of post-consumer vegetable oils.

2.1. Post-consumption vegetable oil

Residual soybean oil, mixture of soybean oil and hydrogenated fat residual: samples of raw materials found in some homes in the city of Campina Grande-Paraíba were collected. These oils had a dark color and unpleasant odor. A sample of fresh vegetable oil from soybeans, Figure 5 was acquired in a business in order to make a comparison with the samples of vegetable oils untreated post-consumer and post-consumer-treated clays under study. Soybean oil was chosen because it is the most widely used in the domestic market and for having little commercial value in relation to other edible vegetable oils, such as olive oil, sunflower oil and corn oil.

Figure 5. Samples of fresh vegetable oils (a) and post-consumer without treatment (b).

2.2. Methodology

Figure 6 shows the flowchart of the methodology used for the bleaching of treaties with calcic bentonite clay, Tonsil and Aporofo, Paraíba region of post-consumer vegetable oils. This method was adapted from de Santos [32] literature.

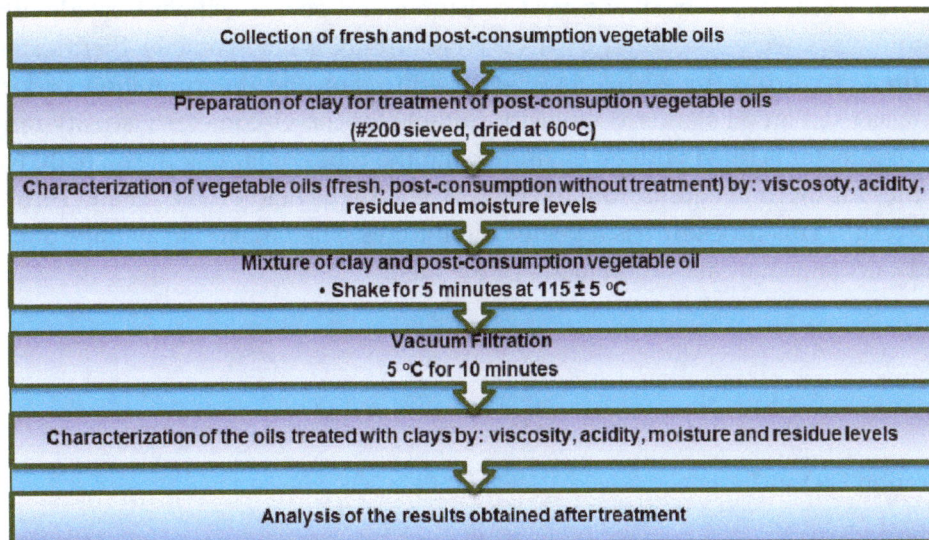

Source: Personal archive.

Figure 6. Steps in the treatment of post-consumer vegetable oils process.

Source: Research data.

Figure 7. Steps used to treat post-consumer vegetable oils: (a) mixing oil with clay, (b) filtration processes the oil with clay, (c) fresh oils and vegetable consumption without post-treatment and (d) post-consumer oil treated and fresh.

2.2.1. Characterization of post-consumer vegetable oils

2.2.1.1. Kinematic viscosity

The kinematic viscosities of fresh oils, post-consumer and post-consumer treated with Tonsil and Aporofo clays were measured by a CANNON-FENSKE viscometer thermostat of the brand Quimis according to ASTM 445, 220V, 40°C. For the determination of kinematic viscosity, was used an oil standard viscosity suitable for the viscometer/viscosity range. The standard viscometer is filled with oil by immersing the tube containing the oil bath, waiting 5 to 10 minutes for thermal equilibrium to occur. A reading is held on the 1st and second bulb, noting the result. This procedure is repeated eight times. Then makes an average of measurements and calculates the calibration factor of the tube:

$$F=\frac{V}{t} \tag{1}$$

F = Calibration Factor

V = Standard viscosity

t = Time spent in seconds

The stirred sample is transferred to a beaker and then the viscometer reservoir is filled with this sample, adapting a stopper at the end, in order to promote complete sealing, thus avoiding leakage of oil. The viscometer is then transferred to a thermostatic bath at a test temperature of 40°C. Then the stopper is removed allowing the flow of oil. It is noted that the time was spent for the oil to drain from the first to second and second to third meniscus viscometer. The kinematic viscosity is calculated by the formula described below, and the result is presented in mm²/s:

$$V_{(cin)}=T\times F \tag{2}$$

where:

V = Kinematic viscosity at the test temperature, in seconds.

T = Time in seconds obtained by the sample flow.

F = calibration factor.

2.2.1.2. Acidity

The acid content of the oils was determined by titration according to ASTM standard D664 24A. For this test, the following reagents are used: isopropyl alcohol (CH₃CH(OH)CH₃);

phenolphthalein indicator ($C_2OH_{14}O_4$); barium hydroxide ($Ba(OH)2.8H_2O$); potassium hydroxide (KOH); potassium hydrogen phthalate ($C_8H_5KO_4$). To prepare the solution of 0.1 NKOH are weighed 5.6g of potassium hydroxide. The solution is transferred to a 1000mL volumetric flask, where it is allowed to stand for 24 hours.

After this time is added 2mL of barium hydroxide to this solution is allowed to stand for 24 hours. Is added 2mL of a solution of barium hydroxide precipitation and if the solution is left standing for 24 hours. If there is no precipitation, the solution is filtered with Millipore filter assembly. Then collects the filtered solution to calculate the factorization of 0.1 N KOH, weighting 0, 3500g of potassium. Hidrogenphthalate is added to the flask 50mL of distilled water and six drops of phenolphthalein indicator. Titrate with 0.1 NKOH It is a white 50mL of distilled water added six drops of phenolphthalein indicator. Titrate again with 0.1 NKOH, recording the volume required. The calculation of the factorization is performed by the following formula:

$$N = \frac{\dfrac{P \times 9,99}{100}}{0,2042 \times (A-B)} \times 56,1 = \tag{3}$$

Where:

P = weight of potassium hydrogen phthalate (grams)

A = Volume of spent KOH titration of potassium hydrogen

B = Volume of spent KOH titration white

V = Normal Concentration

M. Eq. 56.1 KOH =

Purity = 99.9 hidrogenphthalate

Constant = 0, 2042

Constant = 100

In the titration, solvent is titrated in the absence of oil. Initially weighed in an Erlenmeyer ± 2.5g of oil. Added 50mL Erlenmeyer flask with the solvent in the oil, the measuring cylinder and 4 to five drops of phenolphthalein. It is a plug inserted into the Erlenmeyer flask (magnet) for mechanical agitation. Drops of KOH solution are added to the Erlenmeyer flask until the appearance of a slight pink tint. It is noted the amount of KOH. The calculation is done by the NHS expressed below formula and the result is displayed in mg KOH/g.

$$\frac{Factor\,(KOH\,Volume - Volume\,of\,White)}{Weight\,of\,the\,sample} \tag{4}$$

2.2.1.3. Residue

The residue content of the oils was examined in a model centrifuge. 215, brand FANEM, voltage of 220 V. Two tubes are filled with 100 mL sample and then are placed in a centrifuge. The process is centrifuged for 30 minutes at 1,500 rpm.

The result of the residue content is presented in percentage (%).

2.2.1.4. Moisture content

The moisture content of fresh oils, untreated post-consumer and post-consumer treated were analyzed by means of a water condenser with the heating mantle, make Quimis, 220V, Q.321.24 model. Initially, it is checked whether the oil for contamination by water, through the test on a hot plate apparatus. Oil drops are dripped with a glass rod and verifies whether precipitation occurred this oil, i.e., if is detected the presence of water. After this procedure, other tests are initiated to know the amount of oil contamination by water. 0.01mL of sample and 100mL of Xilou into a 500mL flask are added. The water condenser is turned on and starts heating to a temperature of 150°C and adjusted so as to provide reflux for 2 to five drops/second. The process of distillation continued/continues until no more water appears nowhere in the unit, except in the collector. After distillation, the collector is cooled to room temperature. After this process, the reading of the volume of water in the sink is performed. The moisture content is calculated by the following formula:

$$\text{Moisture content } (\%) = \frac{\text{Volume of water in the collector (in mL)} \times 100}{\text{Sample Volume (mL)}} \quad (5)$$

The results of moisture content are presented in %.

Tests of kinematic viscosity, acidity levels, residue and moisture were conducted in the laboratory LUBECLEAN- Distributor Cleansing and Lubricants LTD, located in João Pessoa – Paraíba.

3. Results and discussion

3.1. Analysis of kinematic viscosity, acidity, residue and moisture content

The procedures adopted in this study allowed analyzing comparatively the results Tonsil and Aporofo bentonite clay in viscosity, the levels of acidity, residue and moisture from fresh vegetable oils, untreated post-consumer and post-consumer treaty. Table 1 illustrates these results.

As can be seen in Table 1, the oil treated with Tonsil clay had a lower viscosity compared to the crude oil and indicating the possibility of use as a biofuel is also a better efficiency than the Aporofo clay. The viscosity is a measure of the internal resistance to flow of a liquid, is an important property of vegetable oils because its control feature is intended to maintain its lubricating the engines, as well as proper operation of systems and injection pumps fuel [7]. Values above or below the viscosity specified by the NAP (National Agency of Petroleum, Natural Gas and Biofuels) range can lead to excessive wear on parts self - lubricating the injection system, with an increased work and leak in the fuel pump, as well as providing inadequate fuel atomization, incomplete and with consequent increase in the emission of smoke and particulate matter combustion [42].

	Fresh vegetable oil	Un-treated post consumption vegetable oil	Post consumption vegetable oil treated with Tonsil clay	Post consumption vegetable oil treated with Aporofo clay
Viscosity (mm²/s)	34,57	35,45	32,47	35,33
Acidity (mg KOH/g)	0,64	2,90	2,44	2,27
Residue (%)	0,0	0,1	0,0	0,0
Moisture content (%)	0,0	0,0	0,2	0,3

Table 1. Viscosity values, levels of acidity, and the residue moisture of fresh vegetable oils, untreated post-consumer and post-consumer and treated with Tonsil and Aporofo clay.

Viscosity is a measure of the resistance offered to the flow of diesel engines. The key is to provide a proper atomization of the oil and preserve its lubricating characteristics. Kinematic viscosity of the biodiesel increases with increasing carbon chain and is inversely proportional to the number of unsaturation present thereon. For the same unsaturated compound has a higher viscosity dependence on the configuration of the double bond (cis or trans) than the position of the same [43].

The viscosity directly influences the atomization; that is, the higher the viscosity, the greater the average size of the droplets of the fuel sprays in the combustion chamber. Accordingly, larger droplets resulting in poorer and slower burning mixtures damage the ignition and combustion efficiency. Therefore, an increase in viscosity increases the time delay in the ignition cycle diesel engines.

The viscosity of the biodiesel with increasing carbon chain length and degree of saturation and influences the combustion process in the combustion chamber of the engine. High viscosity causes heterogeneity in the combustion of biodiesel, due to decreased efficiency of atomization in the combustion chamber, causing the deposition of waste in the internal parts of the engine [45].

In Brazil, there was an attempt to make a single specification for biodiesel (B100) similar to those already existing in some countries. However, there are characteristics that differentiate Brazil, a tropical country, countries like those in Europe: the temperature, which is high during most of the year and another important factor, is that in Europe reproduces biodiesel from oils of a single species vegetable, rapeseed. As in Brazil a wide diversity of species of potential use for the production of oils in many cases it is impossible to attain the viscosity values as specified. The specifications for conventional diesel and biodiesel in Brazil are: viscosity at 40°C - NAP 310/01 (diesel) and NAP 255/03 (biodiesel): 2.5-5.5 mm^2/s. The viscosity of the biodiesel is considerably reduced compared to when the source oil is passed to a subsequent process which is the transesterification reaction (which is the step of conversion of the oil or fat to methyl or ethyl esters of fatty acids, which is the biodiesel) [46].

With respect to acid content found in the samples studied after consumption of processed vegetable oils, can be seen a decreased value compared to vegetable oil consumption without post-treatment. NAP Ordinance 42/2004 establishes a value of ≤ 0.80 mgKOH/g for biodiesel (B100) [42]. However, even if the treated vegetable oils do not present results of acidity within the values established by the NAP, these oils must undergo a subsequent treatment process that is the transesterification reaction. The acidity can partly reveal the condition of the oils and fats because the higher the number, the greater the hydrolysis of frying oil, with a consequent increase in fatty acid content. The ideal is that the oil is index less than 2mgKOH/g acidity, in order to have a good reaction yield in the production of biodiesel and also to avoid problems in diesel engines [5].

Argue that the acidity can be defined as the mass (in mg) of potassium hydroxide required to neutralize the free fatty acids not esterified. He also reveals the conservation status of biodiesel, because the hydrolysis of esters occurs with consequent lowering of pH due to the increased content of fatty acids over time [47].

The condition of the oil is closely related to the nature and quality of the raw material, with the quality and purity of the oil with the processing, and especially with the storage conditions. The breakdown of triglycerides is accelerated by heat and light as rancidity is almost always accompanied by the formation of free fatty acid.

The monitoring of acidity in the biodiesel is of great importance during storage, in which the change of values in this period can mean the presence of water. All the rules described above established maximum acidity of 0.5 mg KOHg [45]. The acidity of biodiesel produced, measured in mg KOH per gram of sample, must be strictly within fixed parameters. The National Petroleum Agency recommends an acid, for any sample of biodiesel produced, less than 0.80 mgKOH/g.

A measure of acidity is a variable directly related to the quality of the oils, processing and storage conditions. According to Ordinance No.482 of National Health Surveillance Agency (NHSA 2), the acidity is one of the quality characteristics of various vegetable oils. The lipid materials undergo chemical changes during still in use as heat transfer medium. The acidity is determined by the amount of base required to neutralize the free fatty acids, the acidity increases with the deterioration of the oil during the exchange of heat, such as the frying process, for example [48].

High levels of acidity have a very adverse effect on the oil quality, as to make it unfit for human consumption or even for fuel purposes. Furthermore, the pronounced acidity oils can catalyze intermolecular reactions of triglycerides while affecting the thermal stability of the fuel in the combustion chamber. Also, in the case of use of the fuel oil has a significant free acidity corrosive action on the metallic components of the engine. The acid is a crucial examination for oil and biodiesel since high acidity reaction makes it difficult to produce biodiesel, biodiesel while an acid may cause corrosion of the engine or deterioration of biofuel [8].

An analysis of the results, it is found that soybean oil, despite the relatively high level of acidity, which is a limiting factor of process yield, was higher in the transesterification reaction of the mixture of soybean oil/hydrogenated fat. The yields obtained with either soybean oil as with the mixture, indicate that these materials have significant potential for the production of methyl esters of fatty acids. The positive aspect of biodiesel can be explained by the fact of not having nitrogen or sulfur in structures. Thus, they do not contribute to the acidification of precipitation [13].

The contents of waste clay substantially removed the impurities that were present in the oils without treatment, since the oil recovery achieved by means of a first step which is the bleaching process, is one of the most important steps in refining vegetable oils and has a fundamental role to eliminate substances which color instability and oil (residue). In this step, the decolorization of the oil occurs through the adsorption of pigments, which can be done using clays as adsorbent material [24].

The method of treatment is aimed at reducing the amount of impurities and substances which color the oil. Many of these substances act as catalysts of undesirable reactions, such as oxidation, interfering negatively on the physicochemical characteristics of the oil [24].

With respect to moisture content, all results presented are within the specifications of the NAP. Claim that the oil quality influences the transesterification reaction. So is that the oil is ideal index less than 2.0mgKOH/g of oil and moisture content below 0.5% for the purpose of a good reaction yield in obtaining biodiesel acidity and avoids problems microemulsion, corrosion, among others [49].

With respect to the final destination of clays retained on the filter, these can be applied in layers of compression earthworks or can be converted to non-polluting clusters applicable in the formation of landfills [50].

4. Conclusions

- Tonsil bentonite clay had a potential post-treatment in vegetable oil consumption more efficiently compared with the Aporofo bentonite clay, since clay removed many impurities from the oil.

- The post-consumer oil treated with Tonsil bentonite clay showed a viscosity less than the fresh oil and post-oil consumption without treatment, indicating an excellent adsorptive activity of the clay and the possibility of potential use oil as biofuel.

- Vegetable oils treated clay had a water content lower than after consumption untreated oil acidity.

- The oils treated and characterized for the levels of moisture and residue showed values within the standards established by the NAP.

- The post-consumer vegetable oils showed similar sensory and physicochemical characteristics, but differed substantially from fresh oil.

- After a period of storage, vegetable oils treated post-consumer showed stabilization of properties according to the requirements for processed oil.

- Tonsil bentonite clay may be considered more efficient for the post-consumer treatment vegetable oils since they showed better results than those treated with Aporofo clay.

Acknowledgements

The authors thank the Federal University of Campina Grande, the Graduate Program in Science and Engineering of Materials, the Company LUBECLEAN - Distributor Cleansing and Lubricants LTD, the BENTONISA - Bentonite Northeast S/A and CNPq for fellowships and financial support.

Author details

Elaine Patrícia Araújo[1*], Divânia Ferreira da Silva[1], Shirley Nobrega Cavalcanti[1], Márbara Vilar de Araujo Almeida[2], Edcleide Maria Araújo[1] and Marcus Vinicius Lia Fook[1]

*Address all correspondence to: elainepatriciaaraujo@yahoo.com.br

1 Department of Materials Engineering, Federal University of Campina Grande, Brazil

2 Department of Civil Engineering, Federal University of Campina Grande, Brazil

References

[1] Hocevar, L. Biocombustível de óleos e gorduras residuais - A Realidade do sonho. In: II Congresso Brasileiro de plantas oleaginosas, óleos, gorduras e biodiesel, Varginha, São Paulo, p. 953-957, 2006.

[2] Bhattacharya, A. B., Sajilata, M. G., Tiwari, S. R., Singhal, R. S. Regeneration of thermally polymerized frying oils with adsorbents. Food Chemistry. v.110, p. 562-570, 2008.

[3] Torres, E. A., Chirinos, H. D., Alves, C. T., Santos, D. C., Camelier, L. A. Biodiesel: o combustível para o novo século. Bahia Análise & Dados. Salvador, v. 16, p.89-95, 2006.

[4] Lapuerta, M.; Armas, O.; Ballesteros, R.; Fernández, J.; Diesel emissions from biofuels derived from Sapinsh potential vegetable oils. Fuel, China, v. 84, p.773-780, 2005.

[5] Dantas, H. J., Candeia, R. A., Conceição, M. M., Silva, M. C. D., Santos, I. M. G., Sousa, A. G. Caracterização físico-química e estudo térmico de biodiesel etílico de algodão. http://www.biodiesel.gov.br (accessed 28 August 2008).

[6] Murillo, S., Míguez, J. L., Porteiro, J., Granada, E., Morán, J. C. Performance and exhaust emissions in the use of biodiesel in outboard diesel engines. Fuel, v.86, p. 1765-1771, 2007.

[7] Neto, P.R.C., Rossi, L. F. S., Zagonel, G. F., Ramos, L. P. Produção de biocombustível alternativo ao óleo diesel através da transesterificação de óleo de soja usado em frituras. Química Nova. São Paulo, v.23, 19p. 2000.

[8] Dantas, H. J. Estudo Termoanalítico, Cinético e Reológico de Biodiesel derivado do Óleo de Algodão (*Gossypium hisutum*). 122f. (MS in Química/Química analítica), Centro de Ciências Exaras e da Natureza. Federal University of Paraíba, João Pessoa, 2006.

[9] Graboski, M. S., Mccormick, R. L. Combustion of fat and vegetable oil derived fuels in diesel engines. Prog. Ennerg. Combust. Sci. Colorado, v.24, n.1, p. 57-64, 1998.

[10] Castellanelli, C.A. Study the feasibility of producing biodiesel, obtained through the used frying oil in the town of Santa Maria - RS. p. 1-30. Dissertation (MS in Production Engineering), the Engineering Institute of Paraná. Federal University of Santa Maria, Santa Maria, 2008.

[11] Issariyakul, T., Kulkarni, M. G., Meher, L. C., Dalai, A. K., Bakhshi, N. N. Mixtures of Biodiesel production from canola oil and used cooking oil. Chemical Engineering Journal. Canada, V.140, p.77-85, 2008.

[12] Alberice, R. M.; BRIDGES, F.F.F. Recycling used cooking oil through the manufacture of soap. Environmental Engineering. Espírito Santo do Pinhal, v. 1, n.1, p. 073-076, 2004.

[13] Dorado, M. P., Cross, F., Palomar, J. M., Lopez, F. J. In approach to the economics of two vegetable oil based biofuels in Spain. Renewable Energy. v. 31, p. 1231 - 1237, 2006.

[14] Vecchi, C. C. C., Tarozo, R., Pinto, J. P., Faccione, M., Guedes, C. L. B. Thermal and photochemical degradation process of biodiesel from soybean oil. In: international Congress of r&d in Oil and Gas, 3rd, 2005, Salvador. Proceedings of the 3rd Brazilian Congress of R & D in Oil and Gas, Brazil, 6p, 2005.

[15] Texeira-Neto, E., Teixeira-Neto, A. Chemical modification of clays: scientific and technological challenges for obtaining new products with higher added value. Quim. Nova. São Paulo, v. 32, n. 3, p. 809-817, 2009.

[16] Sanibal, E. A. A., Filho, J. M. Physical, chemical and nutritional oils subjected to frying process changes. Notebook Technology Food & Beverage. Sao Paulo, p. 48-54, 2008.

[17] Vergara, P., Wally, A. P., Pestana, V. R., Bastos, C.; Zambiazi, R. C. Study of the behavior of soyabean oil and rice reused in successive fried potatoes. B. CEPPA, v. 24, p. 207-220, 2006.

[18] AL-Kahtani, H. A. Survey of quality of used frying oils from restaurants. J. Am. Oil Chem. Corporation. Champaign, v.68, n. 11, p. 857-862, 1991.

[19] Corsini, M. S., Jorge, N. Alterações Oxidativas em óleos de algodão, girassol e palma utilizados em frituras de mandioca palito congelada. Alim. Nutr. Araraquara, v.17, p. 25-34, 2006.

[20] O'Brien, R. D. Fats and oils: formulating and processing for applications. Technomic Publishing. Lancaster, p. 385 – 410, 1998.

[21] Tiritan, M. G., Ferreira, E. S. Produção de biocombustível de óleo de fritura com etanol hidratado utilizando planejamento fatorial. In: Anais II Congresso da Rede Brasileira de Tecnologia de Biodiesel, Brasília, 2007.

[22] Silva, C. V. Reaproveitamento do óleo de cozinha como tema nas aulas de educação ambiental. 2010. 48f. (Graduate in Química), Centro de Ciências e Tecnologia, Universidade Estadual da Paraíba, PB, Brasil.

[23] Mothé, C. G., Correia, D. Z. In: Anais do 2° Congresso Brasileiro de Plantas Oleaginosas, Óleos, Gorduras e biodiesel, Varginha, p. 547-551, 2005.

[24] Oliveira, C. G. Proposta de modelagem transiente para a clarificação de óleos vegetais- experimentos cinéticos e simulação do processo industrial. 164p. Dissertação (MS in Engenharia Química), Centro Tecnológico. Universidade Federal de Santa Catarina, Florianópolis, 2001.

[25] Firestone, D., Stier, R. F., Blumenthal, M. M. Journal of Food Technology. p. 90-94, 1991.

[26] Baraúna, O. S. Pigment adsorption process with vegetable oil acid-activated smectite clays.173P. Thesis (Ph.D. in Chemical Engineering), School of Chemical Engineering. State University of Campinas, Campinas, 2006.

[27] Ramalho,V. C., Jorge, N. Ação antioxidante de a-tocoderol e de extrato de alecrim em óleo de soja submetido à termooxidação. Química Nova. São Paulo, v. 29, p. 755-760, 2006.

[28] Cella, R. C. F., Reginato, D. A. B.; Spoto, M. H. F. Ciência e Tecnologia de Alimentos. São Paulo, p.1001-1007, 2005.

[29] Daily Borborema. Biodiesel from frying oil. Campina Grande, December 6, 2008.

[30] Santos, P. S., Science and Technology of Clays. 2 ed. São Paulo: Editora Edgard Blücher Ltda, 1989. 408p.

[31] Dutra, R. P. S., Varela, M. L., Birth, R. M., Gomes, U. U., Paskocimas, C., Melo, P. T. Evaluation of the potential of clays of Rio Grande do Norte - Brazil. Industrial ceramics. São Paulo, v. 11, p. 42-46, 2006.

[32] Santos, P. S. Ciência e Tecnologia de Argilas. 2 ed. São Paulo: Editora Edgard Blücher Ltda., 1992. p. 650-673.

[33] CETEM. Centro de Tecnologia Mineral. Bentonita. Rio de Janeiro, p. 217-230, 2005. (Boletim Técnico,115-00).

[34] Amorim, L. V., Viana, J. D., Farias, K. V., Barbosa, M. I. R., Ferreira, H. C. Comparative study between varieties of bentonite clays of Boa Vista, Paraíba. Matter Magazine. Rio de Janeiro, v. 11, p. 30-40, 2006.

[35] Leite, I. F. Preparação de nanocompósitos do poli (tereftalato de etileno)/Bentonita. 86f. Dissertação (MS em Ciência e Engenharia de Materiais), Centro de Ciências e Tecnologia. Universidade Federal de Campina Grande, Campina Grande, 2006.

[36] Barbosa, R; Araújo, E. M, Melo, T.J.A., Ito, E.N. Preparação de argilas organofílicas e desenvolvimento de nanocompósitos de polietileno. Parte2: Comportamento de Inflamabilidade. Polímeros: Ciência e Tecnologia. São Carlos, v.17, p.104-112, 2007.

[37] DNPM, Departamento Nacional de Produção Mineral, Bentonita, Sumário Mineral Brasileiro, 2007.

[38] Silva, A. R. V., Ferreira, H. C. Argilas bentoníticas: conceitos, estruturas, propriedades, usos industriais, reservas, produção e produtores/fornecedores nacionais e internacionais. REMAP-Revista Eletrônica de Materiais e Processos. Campina Grande, v. 3.2, p. 26-35, 2008.

[39] Barbosa, R. Effect of quaternary ammonium salts on organophilization of national bentonite clay nanocomposites for the development of high density polyethylene (HDPE). p. 7-12. (MS in Chemical Engineering), Centre for Science and Technology. Federal University of Campina Grande, Campina Grande, 2005.

[40] Coelho, A. C. V., Santos, P. S., Santos, H. S. Argilas especiais: argilas quimicamente modificadas - uma revisão. Química Nova. São Paulo, v. 30, p.1282-1294, 2007.

[41] Foletto, E. L., Volzone, C., Morgado, A. F., Porto, L. M. Influência do tipo de ácido usado e da sua concentração na ativação de uma argila bentonítica. Cerâmica. São Paulo, p. 208-211, 2001.

[42] Melo, J. C., Teixeira, J.C., Brito, J. Z, Pacheco, J. G. A., Stragevitch, L. Production of biodiesel from myrtle oil. Biodiesel. Available at: www.periodicosdacapes.com.br. (accessed 10 September 2012).

[43] Knothe, G., Steidley, K.R. Fuel 84 (2005) 1059.

[44] Teixeira, C. V., Colaco, J. M. Viscosity and performance of diesel/biodiesel blends on a single cylinder engine. http://rmct.ime.eb.br/arquivos/RMCT_1_tri_2013/RMCT_007_E4A_11.pdf. (accessed 10 April 2014).

[45] Lobo, I. Ferreira, S. P. Biodiesel: parameters of quality and analytical methods. Quim. Nova, Vol 32, No. 6, 1596-1608, 2009.

[46] Gomes, L. F. S. Potencial de produção de biodiesel a partir do óleo de frango nas co-operativas do Oeste do Paraná. 81f. (MS in Engineering Agrícola), Centro de Ciências Exatas e Tecnológicas. Universidade Estadual do Oeste do Paraná, Cascavel, 2005.

[47] Vasconcelos, A. F. F.; Dantas, M. B.; Lima, A. E. A.; Silva, F. C.; Conceição, M. M.; Santos, I. M. G.; Souza, A. G. Compatibilidade de misturas de biodiesel de diferentes oleaginosas. http://www.biodiesel.gov.br. (accessed 10 March 2013).

[48] Leung, D.Y.C., Guo, Y.Transesterification of used frying neat and oil: optimization for biodiesel production, Fuel Processing Technology, V.87, p. 883-890, 2006.

[49] Santos, R. B. dos, Serrate, J. W., Caliman, L. B., Lacerda, J. R, V.; Castro, E. V. R. Avaliação do uso de óleo residual usado em fritura para a produção de biodiesel e estudo da transesterificação de óleo de soja com álcoois de cadeia de até quatro carbonos. http://www.periodicoscapes.gov.br. (accessed 03 August 2011).

[50] Ramos, A. A. P Study on the recycling process of industrial lubricant level. 67p. (Graduate Mechanical Engineering), Faculty of Engineering. Pontifical Catholic University of Rio Grande do Sul, Porto Alegre, 2008.

9

Issues in the Identification of Smoke in Hyperspectral Satellite Imagery — A Machine Learning Approach

Mark A. Wolters and C.B. Dean

Additional information is available at the end of the chapter

1. Introduction

Observations from earth-orbiting satellites play an important role in the study of various large-scale surface and atmospheric phenomena. In many cases the data collected by such satellites are used and communicated in the form of raster images—three-dimensional data arrays where the first two dimensions define pixels corresponding to spatial coordinates. The third dimension contains one or more image *planes*. A greyscale image, for example, has one image plane, while a color (RGB) image has three planes, one each for the brightness in the red, green, and blue parts of the visible spectrum.

The present work is related to *hyperspectral* images, where the number of image planes is much greater than three. In a hyperspectral image with r planes there is associated with each pixel a set of r data values, each measuring a different part of the electromagnetic spectrum.

The general task of analyzing geographic remote sensing imagery is aptly described by Richards [1] (p. 79):

With few exceptions the reason we record images of the earth in various wavebands is so that we can build up a picture of features on the surface. Sometimes we are interested in particular scientific goals but, even then, our objectives are largely satisfied if we can create a map of what is seen on the surface from the remotely sensed data available...

There are two broad approaches to image interpretation. One depends entirely on the skills of a human analyst—a so-called photointerpreter. The other involves computer assisted methods for analysis, in which various machine algorithms are used to automate what would otherwise be an impossibly tedious task.

Here, we will consider methods that are useful for the second approach: computer-assisted photointerpretation. Computer-aided analysis is particularly helpful for hyperspectral images, which contain too many planes to be visualized in a simple human-readable form.

The present work can be viewed as a case study in the application of machine learning approaches to a difficult task in remote sensing image segmentation. The remainder of this section introduces the problem we are addressing, the data we are using, and the modelling approach we will follow. In Section 2, important ideas from the field of classification are introduced in a tutorial format for researchers who might not be familiar with the topic. Those with prior experience in the area may wish to skip the section. Sections 3 and 4 describe the methods used and the results obtained. Sections 5 and 6 provide discussion and conclusions.

1.1. The problem

The application of interest is the automated identification of smoke from forest fires using hyperspectral satellite images. Smoke released from forest fires can be transported large distances and affect air quality over large areas, making it a matter of population health concern. Despite the importance of smoke events, their spatial scale makes them difficult to quantify through direct measurement. Satellite imagery is an alternative information source that could potentially fill a data gap, providing information about smoke over large areas at times of interest.

The work reported here is the first step in a research stream with the ultimate goal of developing a system that can quantify smoke using moderate- to high-resolution remote sensing images covering large geographic areas, and do so with minimal human intervention. If smoke can be quantified through remote sensing image analysis, the resulting data could be used as input to deterministic predictive models of forest fire smoke dispersal, as a validation check for such models, or as an input to retrospective studies of the health impacts of smoke.

Our present objective is twofold: first, to report our current results in developing a classifier for smoke detection, and second, to stimulate other researchers to consider applying similar methods for their own problems in remote sensing image analysis.

1.2. The data

The region of interest in this study covers parts of western Canada and the northwestern United States, and is centered close to the city of Kelowna, British Columbia. It extends from 46.5° to 53.5° latitude, and from -126.5° to -112.5° longitude. Data come from the moderate resolution imaging spectroradiometer (MODIS) aboard the Terra satellite, which provides images with 36 planes covering different spectral bands ranging from the blue end of the visible spectrum (400 nm) to well into the infrared (14 μ m). More information about MODIS can be found in [3, 4].

The Terra satellite follows a polar orbit that allows MODIS to image most of the globe each day, with images captured at mid-morning local time. All data are freely available from the LAADS web data portal [5]. There are numerous data products available, at different levels of processing for different purposes. We used the Level 1B data at 1km resolution, which provides the hyperspectral data in calibrated form corrected for instrumental effects, but without further manipulation. The data are available in chunks called *granules*. Each granule holds the instrument's observations as it passed over a certain portion of the earth's surface during a particular five-minute time interval. If a study region does not happen to be covered by a single granule, it is possible to stitch the data from adjacent granules to cover the region. If the region is large enough, it may be necessary to stitch granules from different orbital passes. In our case, we only used data from time-sequential granules, and not those from different passes, because we found that the smoke and clouds in the scene could change significantly between orbital passes. Because of this it was not always possible to collect complete data for the entire region of interest on every day.

A total of 143 images were collected, one for each day covering the peak dates of the fire season (July 15 to August 31) for the years 2009, 2010, and 2012. Each image is approximately 1.2 megapixels in size, and has spatial resolution of approximately one kilometer per pixel. Images are in plate carrée projection. Any pixel that had data quality concerns (as indicated by error codes in the downloaded data) was excluded from the analysis. The entirety of band 29 was also discarded because of a known hardware failure, leaving 35 spectral bands to be used for classification purposes.

To aid in visualization of the data, an RGB version of each image was produced. Following [6], the RGB images were created by letting bands 1, 4, and 3 fill the red, green, and blue image planes, respectively. First, each of these three bands was run through a saturating linear brightness re-mapping, letting 1 percent of the pixels be saturated at each end of the brightness range. Then, a piecewise linear brightness transformation was carried out on each band, as in the reference.

The resulting RGB images were used for the important task of manually assigning each pixel to either the smoke or nonsmoke class—that is, for specifying what the "true class" of each pixel was. To make this task easier, fire locations (found by comparing bands 22 and 31, as in [7]) were overlaid on the RGB images. While the smoke was sometimes easy to distinguish from the rest of the image, there were also many cases where the choice of true class was quite ambiguous: regions where smoke and cloud were mixed, or regions where the smoke was not highly concentrated, for example. Nevertheless, each pixel in all 143 images was assigned a true class label on a best-efforts basis. The approach to assigning true labels was to assign the smoke class whenever a pixel appeared to have any level of smoke, even a thin haze. The end result was a set of 143 black and white *mask* images corresponding to the hyperspectral ones, with white pixels indicating smoke and black indicating nonsmoke. The complete set of masks comprised 90% nonsmoke pixels and 10% smoke pixels.

As will be shown at the end of this chapter, the difficulty assigning true classes with high confidence is a potentially critical limitation of the analysis. The manual approach to labelling was used nonetheless, since no alternative method exists for identifying smoke pixels across entire images. We note in passing that we have previously obtained some "gold standard" images by request from NASA, and in this case smoke was also identified as hand-drawn regions.

1.3. Modelling approach

The observed images are the product of natural processes that are very complex. From a statistical standpoint, a sequence of remote sensing images covering a particular region of the earth is a spatiotemporal data set with statistical dependence both within and between images. Physically, the presence of smoke in a particular region at a particular time is surely dependent on the characteristics of a particular fire, as well as on meteorological and topographical variables that vary over the region of interest and over time. There is thus ample scope for mathematical complexity in a model used for classification. Some decisions must be made at the outset about which aspects of the problem to include in our classifiers, and which to ignore. As the research is still in its early stages, three simplifying decisions have been made.

First, classification will be conducted based only on the spectral information in the images themselves; no ancillary information (for example, about wind, fire locations, or topography) will be used to aid prediction. This decision was made partly to limit model complexity, but also to ensure that our methods are wholly independent of any physics-based deterministic models (which they might eventually be used to validate). Using only the hyperspectral data also maximizes the applicability of the methods to other image processing tasks.

Second, the focus is on detecting only the presence or absence of smoke. A successful system will be able to classify images on a pixel-by-pixel basis into one of two categories, "smoke" or "nonsmoke."

Third, all pixels and all images are assumed to be independent of one another. While ignoring temporal dependence from image to image does not throw away much information—with images collected at a frequency of once per day, there is little correlation between smoke locations from one image to the next—ignoring spatial dependence within images is clearly making a compromise. Smoke appears in spatially contiguous regions, so knowledge that a certain pixel contains smoke should influence adjacent pixels' probability of being smoke. Nevertheless, spatial association between the outcomes introduces many technical difficulties, so it was not included at this stage of our study.

With these decisions, the smoke detection task becomes a typical *binary classification* or *binary image segmentation* problem, using the data in the 35 spectral bands as predictors. Simplifying the problem in this way is justified in a preliminary analysis. Our goal is to evaluate whether the spectral data contain enough information to allow the smoke and nonsmoke pixels to be distinguished from one another with reasonably high probability. If they do not, there is little to be gained from the added complexity of more sophisticated models; if they do, the simple independent-pixel smoke/nonsmoke model can be extended in a variety of ways to obtain further improvements. Furthermore, it will be seen that despite retreating to a simple model for classification, the problem is still high dimensional, computationally intensive, and challenging.

With these considerations in mind, we use logistic regression for building our classifiers. Logistic regression has convenient extensions for accommodating spatial associations, for handling multiple levels of smoke abundance, and for including additional predictor variables. We anticipate that a final, useful future system will be based on such an extended model.

All analyses presented here were carried out using the free and open source statistical computing software R [2]. An R script demonstrating much of the analysis is available on the corresponding author's website (www.mwolters.com); readers interested in working with the full data set (which is large) can contact the authors by email.

2. Binary classification concepts

Classification is the process of assigning a category (a class label) to an item, using available information about the item. We are interested in binary classification, where there are only two class labels. In our case, the labels are nonsmoke (class 0) and smoke (class 1), the items to be classified are image pixels, and the available information is the content of the hyperspectral image. We say we have "built a classifier" when we have established a rule that tells us how any given pixel in a new image should be classified.

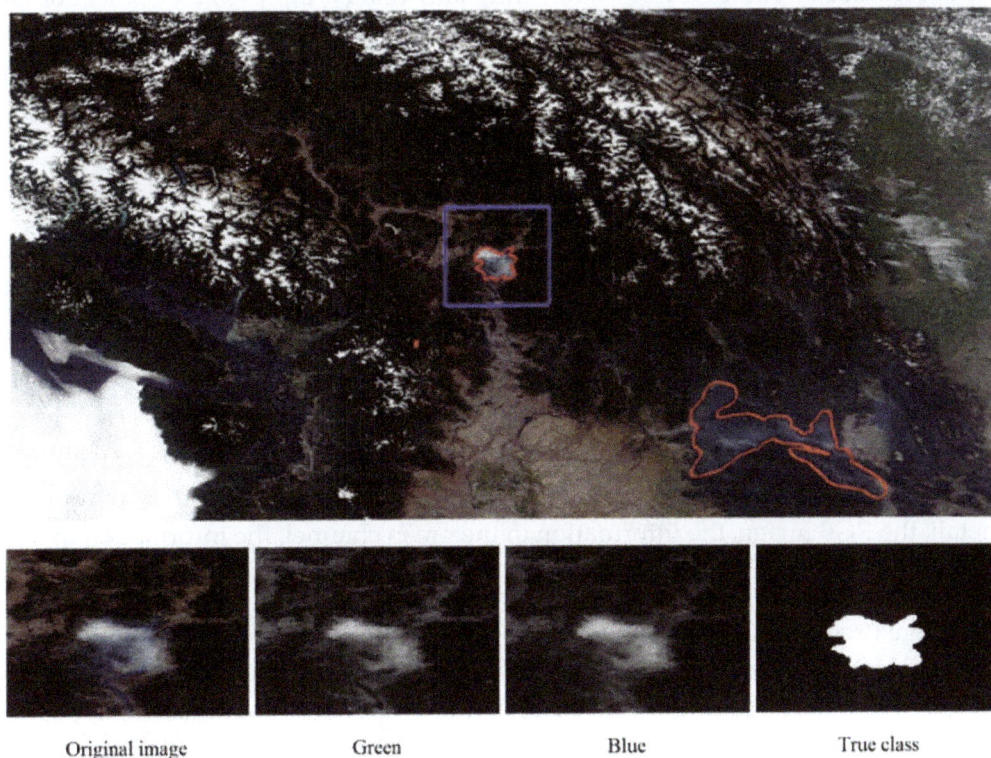

Original image　　　　　Green　　　　　Blue　　　　　True class

Figure 1. The data used for the example. Top: an RGB image of the study region, with regions of smoke outlined in red. The blue rectangle encloses the pixels that are used for the example. Bottom, from left to right: the RGB sub-image of the region of interest, the green channel, the blue channel, and the mask showing the true smoke (white) and true nonsmoke (black) regions.

Classifier building requires the availability of *training data*—a set of items where the true class labels are known. The reliance on training data is one reason classification is also known as *supervised learning.* One may think of an all-knowing supervisor who tells us the class membership of a subset of our items, but then goes home for the day, leaving us to learn for ourselves how to classify the remaining items. To prevent confusion, note that the alternative problem of *unsupervised learning* (where the wise supervisor never shows up, leaving all class labels unknown) is also known as *clustering*, and—although important in its own right—is not presently relevant.

Classification is a large topic. It is, in fact, the dominant activity in the field of machine learning. Consequently, no attempt is made here to provide a thorough review of the subject. Rather, a single classifier based on logistic regression will be discussed as a means of introducing common themes in classification. The logistic classifier is naturally suited to binary classification problems, and has a relatively simple form with strong connections to linear and nonlinear regression. This classifier will be used throughout the chapter.

Readers interested in further background on classification, and alternative classifiers, have many resources to turn to. The books [1, 8, 9, 10] provide accessible introductions to the topic, and [1] in particular discusses classification and many related topics in the context of remote sensing imagery. Note that while alternative classification methods may have better or worse performance in different situations, most of the important aspects of setting up and solving a classification problem remain the same regardless of the particular method chosen.

2.1. A small example

As an illustrative example, we restrict our attention to a small subset of the study data—a portion of a single image—and work with only the RGB image rather than the full hyperspectral data. The large image in Figure 1 shows the entire study region on the chosen date (and also provides an example of what the color images look like on a clear day). The picture contains two areas outlined in red. These are the areas that were deemed to contain smoke during the masking process. The blue rectangle in the image outlines the set of pixels used for this example. The four smaller images at the bottom of the figure show the example data in more detail: the RGB image, the information in the green channel, the information in the blue channel, and the corresponding mask showing the true classes.

The sub-image used for the example is 150 by 165 pixels (24750 pixels in all) and is centered on a smoke plume. To allow the problem to be visualized in two dimensions, we will consider only the green channel (G) and the blue channel (B) as predictors in our classifier.

2.1.1. Logistic classifier with two predictors

The logistic classifier is based on logistic regression, which is set up as follows. Let the true class (the response variable) of the i th pixel be Y_i, with $Y_i = 1$ corresponding to smoke and $Y_i = 0$ corresponding to nonsmoke. The true class is modelled as a Bernoulli random variable with $\pi_i = P(Y_i = 1)$ being the probability of the smoke outcome. All pixels are assumed to be statistically independent.

Logistic regression models the log-odds of pixel i being smoke (the event $Y_i=1$) as a linear combination of predictor variables (the green and blue brightness values, in this case):

$$\log\left(\frac{\pi_i}{1-\pi_i}\right) = \beta_0 + \beta_1 G_i + \beta_2 B_i, \tag{1}$$

where G_i and B_i are the green and blue values of the i th pixel, and $\{\beta_0, \beta_1, \beta_2\}$ are the model coefficients. These three coefficients are to be estimated from a set of pixels for which both the responses and the predictors are known. Estimation is done using a weighted least squares or (equivalently) maximum likelihood approach. The process is called *model fitting* or *training*, and software for performing the estimation is readily available.

Once the parameters are estimated, the fitted model can be used to generate predictions for any given pixel, whether or not the response has been observed. Let x_j represent such a pixel, with predictor values G_j and B_j. Plugging G_j, B_j, and the fitted coefficients into the right hand side of (1), the equation can be solved for $\hat{\pi}_j$, the *fitted probability*. This quantity is the estimated probability that pixel j belongs to the smoke class.

The logistic regression model gives us fitted probabilities on a continuous scale from zero to one. To convert the model into a binary classifier, one need only specify a cutoff probability, c. If $\hat{\pi}_j$ is less than c, pixel j will be put into class 0 (nonsmoke), and if $\hat{\pi}_j$ is greater than c, it will be put into class 1 (smoke). We choose $c=0.5$, so that each pixel is put into the class that is more probable under the model.

Returning to the example data, the above procedure was followed using the 24750 chosen pixels and their true class labels as training data to fit model (1). The nature of the resulting fitted model is shown in Figure 2. The figure plots each pixel as a point in the (green, blue) plane. In machine learning, predictor variables are often called *features*, and so this plot considers each pixel in the model's *feature space*. We see that the smoke pixels generally occur at higher values of both blue and green, but that there is overlap between the two classes; the two classes are not completely separable. The fitted logistic regression model allows us to calculate a probability of being smoke for any point in the feature space. The thick line on the plot is the probability 0.5 contour of this probability surface; it is the decision boundary for our classifier with $c=0.5$. The model will classify any pixel above this line as smoke, and any pixel below the line as nonsmoke.

The inset image in the figure shows the classifier's predictions. White pixels in this image indicate pixels estimated to have greater than 50% chance of being smoke. The red outline indicates the boundary of the true smoke region. While most of the pixels are classified correctly, many are not.

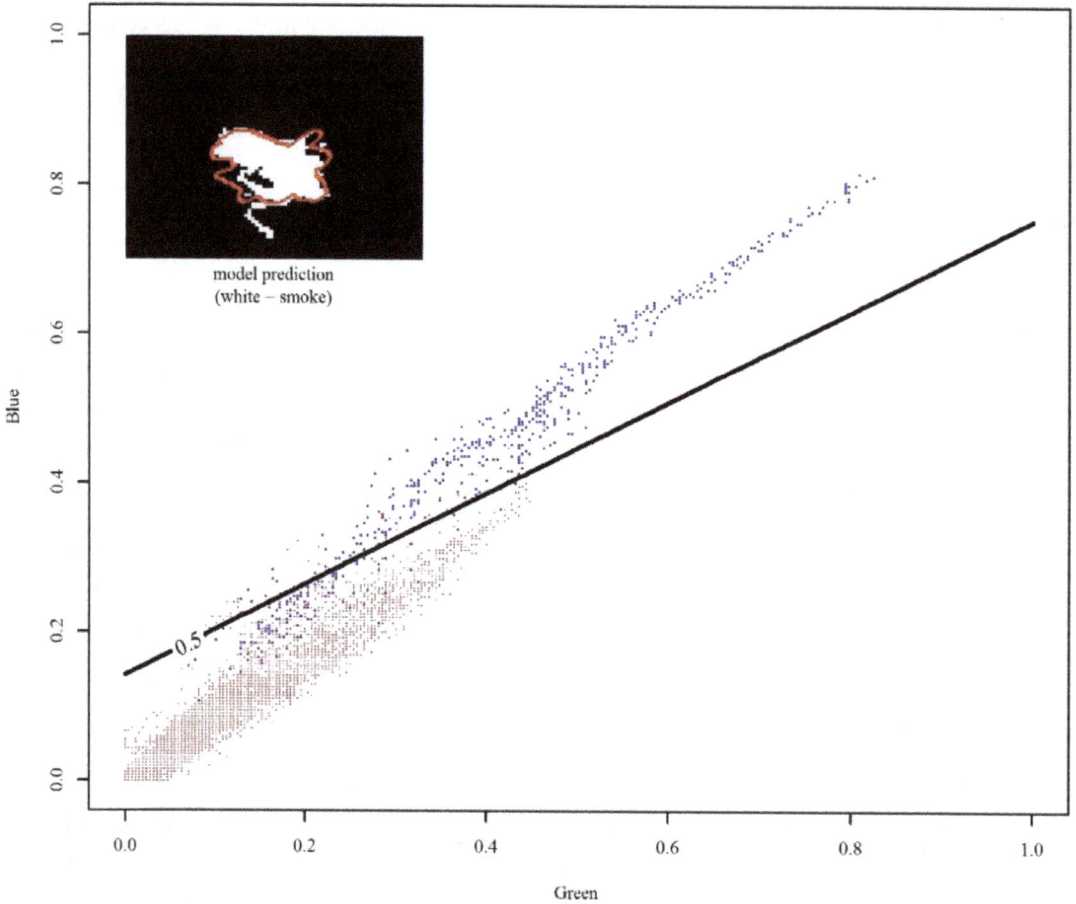

Figure 2. Results of fitting the two-predictor model (G, B) to the example image. Blue points are smoke pixels and red points are nonsmoke. The line on the plot gives the 50% probability line that can be used to discriminate one class from the other. The inset image shows the predicted classes using this model; the red outline in the inset is the boundary of the true smoke region.

2.1.2. Logistic classifier with expanded feature space

The mathematical structure of the previous model ensured that the decision boundary in Figure 1 had to be a straight line. This limited the ability of the classifier to discriminate between the two classes. To make the model more flexible, we can expand the size of the feature space by adding nonlinear functions of the original predictors G and B. For example, we can consider the model

$$
\begin{aligned}
\log\left(\frac{\pi_i}{1-\pi_i}\right) = \beta_0 &+ \beta_1 G_i + \beta_2 B_i, + \beta_3 G_i^2 + \beta_4 B_i^2 + \\
&+ \beta_5 G_i B_i + \beta_6 G_i^3 + \beta_7 G_i B_i^2 + \beta_8 B_i G_i^2 + \beta_9 B_i^3 + \beta_{10} G_i^2 B_i^2,
\end{aligned}
\tag{2}
$$

which includes the original variables G_i and B_i, along with squared and cubed terms (like G_i^2 and G_i^3) as well as products between the original variables taken to various powers (as in $G_i B_i$ and $B_i G_i^2$). Borrowing terminology from industrial experimentation, we call the original variables *main effects* and any terms involving products of variables *interactions*.

The right hand side of model (2) is still a linear combination of various predictor variables, but we have expanded the feature space to ten dimensions. Considered as a function of G and B, the model is able to handle nonlinear relationships between these main effects. In Figure 3 we see the results of fitting this model to the example data. The figure shows the same scatter plot of the data, but now with the 50% contour line for this more flexible model. By adding extra features we can define a decision boundary with more complex shape. The additional shape flexibility of this boundary allows the classifier to correctly assign classes to a greater proportion of the pixels, as seen in the inset prediction image.

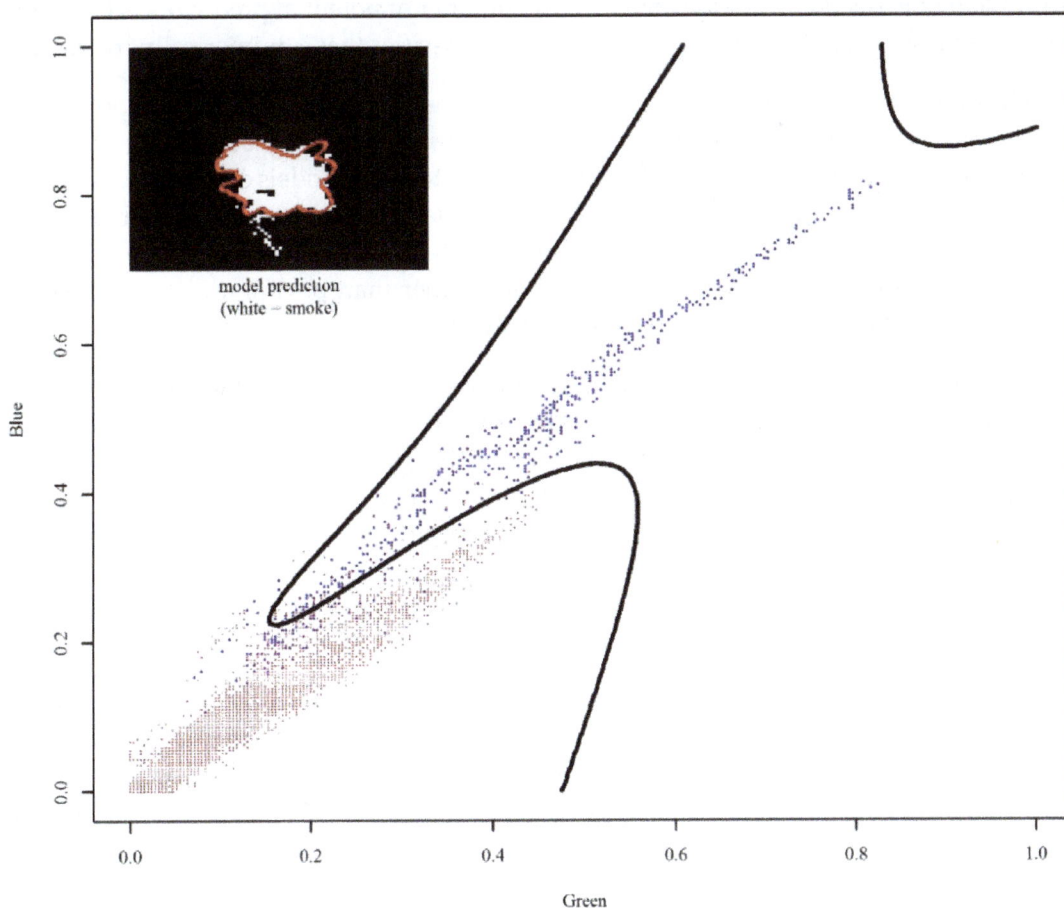

Figure 3. Results of fitting the example data to the 10-predictor model (G, B, G^2, B^2, GB, G^3, GB^2, BG^2, B^3, G^2B^2). The plot is constructed in the same way as the previous figure. In this case the class decision boundary can take a complex non-linear shape.

2.2. Other important concepts

The preceding example might tempt one to believe that simply adding more predictors to the model will always yield a better classifier. This is not true, however, for two reasons.

The first problem with arbitrarily growing the feature space is purely computational. In most problems (and certainly in the present study), the measured main effects are correlated with each other to varying degrees. When expanding the feature space, the variables in the model will increasingly suffer from a form of redundancy known *multicollinearity*: certain predictors can (almost) be written as linear combinations of the other predictors. When the degree of multicollinearity is mild, model fitting will still be possible, but the coefficient estimates can be grossly inaccurate (and can vary greatly from sample to sample). As the problem gets worse, fitting will fail due to the occurrence of numerically singular matrices in the estimation routine.

The multicollinearity problem does not preclude us from considering a large feature space, but it means we cannot include *all* variables from a large feature space in the model. This leads to the problem of *model (feature) selection*: when the number of potential predictors is large, we seek to choose a subset of them that produces a good classifier that is numerically tractable.

When selecting a model from a large collection of correlated predictors, it is important to remember that the coefficient estimate of a particular variable will vary depending on which other variables are included in the model. Further, the best-fitting models of two different sizes need not share their variables in common (the variables selected in the best five-variable model, for example, might not be present in the best ten-variable model). For these reasons it is best to consider the performance of a model as a whole, rather than paying undue attention to coefficient values, statistical significance tests, and the like.

The second problem is more fundamental, and can arise even when multicollinearity is not present. The predictions shown in the previous figures were predictions made on the training data itself; the same data were used both for model fitting and for evaluating performance. This circumstance leads to *overfitting* and poor *generalization* ability: the model fits the training data very well but, because the training data is only a sample from the population, the model's predictive power on new data suffers. When considering increasingly complex models, a point is reached at which additional complexity only detracts from out-of-sample prediction accuracy.

The remedy for overfitting again involves model selection. Because of overfitting, larger models are not necessarily better, so the challenge is to select a model of intermediate size that is best at what is really important, out-of-sample prediction. To do this, one must use different samples of the data for different parts of the procedure. Ideally, one portion of the data (a training set) is used for fitting, another portion (a *validation set*) for model selection, and a third portion (a *test set*) for final evaluation of predictive performance ([9], p. 222).

A final important consideration is the particular measure used for evaluating classifier performance. Any item processed by a binary classifier falls into one of four groups, defined by its true class (0 or 1) and its predicted class (0 or 1). The rates of these four outcomes can be displayed in a so-called confusion matrix, as shown in Table 1. The values a, b, c, d in the table are the rates (relative frequencies) of the four possible outcomes. They must sum to 1. The values b and c (shown in bold) are the rates of the two types of errors: nonsmoke classified as smoke, and smoke classified as nonsmoke. The row sums f_0 and f_1 are the true proportions of items in each class.

Three error rates derived from the confusion matrix are considered subsequently. The *overall error rate* (OER$=b+c$) is simply the global proportion of pixels misclassified. The *classwise error rates* are the rates of misclassification in each class considered separately. We denote these by CER0$=b/f_0$ for the nonsmoke class, and CER1$=c/f_1$ for the smoke class.

Minimizing the OER will be taken as the primary goal of classifier construction. Note however, that our data set consists of 90% nonsmoke pixels ($f_0=0.9$), so focusing on overall prediction performance implicitly puts more weight on prediction accuracy in the nonsmoke class. Because the data are so unbalanced, even the naïve classification rule "assign all pixels to class 0" can achieve an error rate of only 10% (OER$=0.1$), but with the highly unsatisfactory classwise rates CER0$=0$ and CER1$=1$. More will be said about the trade-off between OER and CER in later discussion.

		Prediction		
		Class 0	Class 1	Sum
Truth	Class 0	a	b	$a+b=f_0$
	Class 1	c	d	$c+d=f_1$
	Sum	$a+c$	$b+d$	1

Table 1. A confusion matrix. Values in bold represent errors.

3. Experimental methods

The methods just described were applied to the full set of hyperspectral data. The logistic regression classifier was used, just as in the example. In the full-scale analysis, however, it was necessary to handle a much larger data set and a much larger pool of predictor variables. The following sections describe the methods used for preparing the data and searching for a suitable classifier.

3.1. Data splitting and sampling

This analysis took place in a data-rich context. Having a high volume of data is very advantageous, since the available pixels can be split into separate training, validation, and test groups with each group still having more than enough pixels to yield good estimates of the various quantities of interest. The data were randomly split into these three groups at the image level, with a roughly 50/25/25% split: 70 images (82×10^6 pixels) for training, 36 images (42×10^6 pixels) for validation, and 37 images (43×10^6 pixels) for testing.

The drawback of having this much data is the level of computational resources required to handle it. Fitting the logistic regression model requires matrix computations that are memory and computation intensive when the number of cases (pixels) or the number of predictors become large. To estimate a model with the 35 spectral bands as predictors using the full set of training images, for example, approximately 23 GB of RAM is be required just to hold the data in memory. Special techniques are required to perform regression computations on data sets this large. Furthermore, it is necessary to perform model fitting iteratively as part of a model search step, so simple feasibility is not sufficient. Computational run time is also an important factor.

A practical approach to working with such large data sets is to randomly sample a manageable subset of the data, and work with the sample instead. This approach will work well if the sample size can be chosen such that the computations are feasible and sufficiently fast, while still providing estimates of needed quantities (coefficient estimates, prediction error rates) that are sufficiently accurate.

To determine whether such a sample size could be found in the present case, a sequence of preliminary trials was carried out on the test and validation images. In these trials, the model with 35 main effects was fit to numerous independent training samples, and predictions were made on numerous independent validation samples. It was found that sampling 10^5 pixels was adequate for both the training and validation data. At this sample size, predicted probabilities from fitted models exhibited only minor variations (typically differing less than 0.02) when computed from different samples. Similarly, when the validation sample was this size, estimates of prediction error had variance low enough that it should be possible to estimate the prediction error rate on the full validation set to better than the nearest percentage point.

A working sample of 10^5 pixels was therefore drawn from the test images, and an equal-sized sample was drawn from the validation images. Subsequently all parameter estimation and model selection was done using these two samples, rather than the original images.

3.2. Model families considered

In an attempt to build a successful classifier, four groups of models were considered. Each group was defined by i) the set of candidate predictors that have the opportunity to be selected in the model, and ii) the methods used for model selection and model fitting. We attempted to find a single "best" classifier within each group, and carried forward those four best models for subsequent performance evaluations.

Scenario 1: *RGB model*. This model was the same as the first classifier shown in the earlier example, except with all three variables (R, G, B) used instead of only two. This model was included only as a reference point, since it was not expected to perform particularly well. There is only one possible model in this group, so no model selection step was necessary. Coefficients were estimated in the usual least-squares manner for logistic regression.

Scenario 2: *main effects model*. This model family used the 35 hyperspectral bands as candidate predictors. An optimal model with 35 or fewer variables was to be chosen by subset selection. Coefficients were estimated by least squares.

Scenario 3: *all effects model (subset selection)*. The third set of models included a greatly expanded set of predictors. The complete set of candidate variables for this case includes the following sets of variables:

- All 35 main effects.
- The 35 square-root terms.
- The 35 squared terms.
- The 595 interactions between different main effects.
- The 595 interactions between different square-root terms.
- The 595 interactions between different squared terms.
- The 1225 interactions between main effects and square-root terms.
- The 1225 interactions between main effects and squared terms.

In all, there are 4340 candidate variables in this collection. A best model consisting of a (relatively) small portion of these variables was found by subset selection, and coefficient estimation was done by least squares.

Scenario 4: *all effects model (LASSO selection)*. The fourth group of models used the same set of 4340 candidate predictors, but with model selection and parameter estimation carried out using the LASSO technique. Briefly, LASSO is a so-called *shrinkage* or *regularization* method, where parameter estimation and variable selection are done simultaneously. It works by introducing a penalty term into the least squares objective function used to fit the model. The nature of the penalty is such that certain coefficients are forced to take the value zero, effectively eliminating the corresponding variables from the model. The size of the penalty is controlled by a parameter; the larger this parameter, the more variables are removed from the model. The reader is referred to the literature for further details on LASSO and other shrinkage methods (for example, [11, 12, 9]). The LASSO-regularized logistic regression classifier was constructed using the R package glmnet [13].

3.3. Model selection

The main effects and all effects models required model selection by *best subsets*. For a given set of candidate predictors, this approach to model selection depends on two things: an objective function defining how "good" a particular model is, and a search procedure for finding the best model among all possibilities.

In the present case we were interested in out-of-sample prediction performance, so we used the validation sample of pixels to measure the quality of any proposed model. A straightforward measure of model quality is the prediction error rate on the validation data. While this measure could have been used, here a quantity known as *deviance* was used instead. The deviance is defined as -2 times the log-likelihood of the data under the model, and can be interpreted as a measure of lack of fit (smaller deviance indicates a better fit). For the logistic regression model with n pixels, the deviance is

$$-2\sum_{i=1}^{n}d_{i}, \quad \text{where} \quad d_{i} = \begin{cases} \log(\hat{\pi}_{i}) \text{ if pixel } i \text{ is smoke} \\ \log(1-\hat{\pi}_{i}) \text{ if pixel } i \text{ is nonsmoke,} \end{cases} \tag{3}$$

where $\hat{\pi}_{i}$ is the predicted probability of pixel i being in class 1. We can see from the equation that the i th pixel's deviance contribution, d_{i}, shrinks to zero when the predicted probability gets closer to the truth (i.e., when a smoke pixel's predicted probability approaches one, or when a nonsmoke pixel's predicted probability approaches zero). An advantage of the deviance is that it depends in a smooth and continuous way on the fitted probabilities, whereas the prediction error depends only on whether the $\hat{\pi}_{i}$ values are greater or less than the cutoff c.

In best subsets search, then, the objective function value for any proposed model was found by first estimating the model's coefficients using the training data, and then computing the deviance of the fitted model on the validation data.

Having defined an objective function, it was necessary to search through all possible models to find the best (i.e., minimum deviance) one. This task is challenging, because the combinatorial nature of subset selection causes the number of possible models to grow very quickly when the number of candidate predictors becomes large.

Let the *size* of a particular model be the number of predictors in the model, not including the intercept. Denote model size by k. For the main effects scenario with 35 predictors, there are a manageable 6454 possible models when $k=3$ (i.e., there are 6454 combinations of 3 taken from 35). When $k=5$, however, there are about 325 thousand models from which to choose; and when $k=15$, there are 3.2 billion models. For the all effects scenario with 4340 predictors, the situation is naturally much worse. Even for models of size 3, there are about 13.6 billion possible choices. For larger values of k, the number of possible models becomes truly astronomical, with approximately 10^{30} ten-variable models and about 10^{154} 70-variable models.

Clearly, it is not feasible to search exhaustively through all possible models for either the main effects or all effects scenario. Rather, a search heuristic is required to find a good solution in reasonable time. A traditional approach in such cases is to use sequential model-building procedures like forward, backward, or stepwise selection [14]. These methods have the advantage of convenience, but they lack a valid statistical basis and are generally outperformed by more modern alternatives.

An alternative option, that was pursued here, is to use a more advanced search heuristic to search the space of possible models. We used the function kofnGA, from the R package of the same name [15], to conduct model search using a genetic algorithm (GA). This function searches for best subsets of a specified size, using a user-specified objective function (which we chose to be the validation-set deviance). Instead of considering all possible model sizes, separate searches were run at a range of chosen k values. These were:

For the main effects model: $k=3, \quad 5, \quad 10, \quad 15, \quad 20, \quad 25, \quad 30$.

For the all effects model: $k=3, \quad 10, \quad 20, \quad 30, \quad 40, \quad 50, \quad 60, \quad 70$.

By running the search at only these sizes, we expected to find a model close to the optimal size, without requiring excessive computation times. A discussion of GA methods is beyond the scope of this work, but references such as [16, 17, 18, 19] can be consulted for further information.

When using a search heuristic like GA on a large problem like this, we do not expect that the search will result in finding the single globally-optimal model in the candidate set. In fact if we were to run the search multiple times, it is likely that a variety of solutions will be returned. Nevertheless, the GA can be expected to find a good solution — that is, one with a validation-set deviance close to the minimum — in reasonable time. In practice we expect any model near the minimum deviance will have nearly equivalent predictive performance.

The model selection in the LASSO scenario was done quite differently. As mentioned previously, the LASSO solution depends on a regularization parameter that controls the complexity of the fitted model. For any given value of this parameter, a single model results, with some coefficients zero and some nonzero — the size of the model is implicit in the solution, and is not directly controlled. Model selection thus involves choosing only the value of the regularization parameter. Following the advice of [13], we used validation-set deviance as the measure of model quality for the LASSO fit, and chose the regularization parameter to minimize this quantity.

Note that the LASSO approach enjoys a computational efficiency advantage over the GA-based subset selection approach. For our large training and validation samples (10^5 pixels), fitting the LASSO at 100 values of the regularization parameter took approximately two hours on a contemporary desktop system, while a the longer GA runs (say, with all effects and $k = 50$) took an entire day. Given the overall timeframe of a study like this one, however, the run time difference is not viewed as especially important.

3.4. Performance evaluation

Predictive performance of the best models selected from each group was measured by the overall and classwise error rates OER, CER0, and CER1, as defined in Section 2.2. The probability cutoff c used to map the fitted probabilities onto the two classes was set to its default value of 0.5 for this performance comparison. There is no guarantee that 0.5 actually provides the best value, however. To investigate the impact of varying c, performance of the best model in group 3 was evaluated at a range of c values.

As an adjunct to quantitative assessment, qualitative analysis of model predictions was carried out by visual inspection of the predicted probability maps — greyscale images in which the intensity range [0, 1] represents the predicted probability of each pixel being smoke — from the best model in group 3. For all 37 test images, the probability maps were compared to the original RGB images, to learn more about which aspects of smoke detection were done well, and which were done poorly.

4. Results

The data splitting, sampling, and model selection procedures just described were carried out on the study data, with the net result of producing one best classifier from each of the four scenarios. These four best classifiers were subsequently used to generate predictions for every pixel in the 37 test images. The results of these tasks are presented below, beginning with model selection, and then moving on to the quantitative assessment of prediction performance. The qualitative assessment of performance is reviewed in Section 5.

4.1. Model selection results

The results of model selection are shown in Table 2 and Table 3. The first table lists all of the models considered, along with their deviance and their error rates on the validation data. The error rate estimates in the table are preliminary only, because they are measured on the same validation sample that was used to do variable selection. The final and most accurate measure of out-of-sample predictive performance (the error rates on the test images) are reported in the next section.

The four models selected as best in the four groups are shown in bold in Table 2. For model 1 (RGB), there was only one model, which was selected best by default. For models 2 and 3 (the main effects and all effects models), the best models had $k=20$ and $k=50$, respectively. For model 4 (the LASSO), the minimum-deviance approach chose a model with 109 variables.

Scenario/Model	Results on VALIDATION sample			
	Deviance	OER (%)	CER0 (%)	CER1 (%)
1. RGB	**58549**	**10.2**	**4.3**	**98.4**
2. Main effects, k variables				
$k=3$	57245	10.0	0.0	100.0
$k=5$	53162	9.8	0.4	94.3
$k=10$	50399	9.3	0.5	87.9
$k=15$	48521	8.7	0.5	83.0
$k=20$	**48483**	**8.6**	**0.5**	**82.1**
$k=25$	48704	8.8	0.6	82.8
$k=30$	50144	8.8	0.7	81.7
3. All effects, k variables				
$k=3$	51262	9.6	0.4	92.9
$k=10$	42442	7.6	1.1	65.9
$k=20$	40180	7.2	1.1	62.2
$k=30$	39785	7.1	1.2	60.0
$k=40$	38600	6.8	1.3	55.7
$k=50$	**38174**	**6.8**	**1.4**	**56.0**
$k=60$	38424	6.9	1.6	54.5
$k=70$	38475	6.8	1.6	53.7
4. All effects, LASSO*	**47711**	**8.1**	**1.6**	**66.6**

*The LASSO model shown is the minimum-deviance one, which had 109 nonzero coefficients.

Table 2. List of models considered, with results for the validation set.

Table 3 shows the particular combinations of variables that were chosen in the best models from each of the four groups. The main-effects-only model had 20 variables, the all-effects model had 50 variables, and the LASSO model had 109 variables (of which only 50 are shown). When regression models become this large, it is very difficult to glean any useful information from lists of included variables. Nevertheless, the table is presented for the sake of completeness.

---

1. RGB Image:

Red, Green, Blue (nonlinear transformations of bands 1, 4, and 3)

2. Main effects, $k=20$:

21, 31, 32, 24, 25, 36, 18, 7, 1, 23, 17, 6, 8, 30, 13, 11, 14, 16, 26, 15

3. All effects, $k=50$:

19:$\overline{21}$, 26:$\overline{21}$, 24:$\overline{26}$, $\overline{3}$:28, 5:$\overline{24}$, 23:$\overline{1}$, 28:$\overline{4}$, 2:$\overline{25}$, 33:$\overline{21}$,

$\overline{30}$:7, 8:$\overline{25}$, 27:$\overline{31}$, 30:$\underline{20}$, 22:$\underline{2}$, 8:23, 31:$\overline{30}$,

$\underline{9}$:$\overline{18}$, 24:$\overline{27}$, $\underline{23}$:8, 30:$\underline{34}$, 27:$\overline{23}$, 8:$\underline{32}$, 11:$\overline{36}$,

11:36, $\overline{4}$:$\underline{9}$, 16:$\underline{19}$, 23:$\overline{19}$, 27:$\overline{36}$, 7:$\underline{23}$, 11:6,

$\overline{16}$:17, 5:$\underline{1}$, 20:$\underline{14}$, $\underline{19}$:$\underline{25}$, 36:$\overline{35}$, $\underline{7}$, $\overline{16}$:17,

20:$\overline{34}$, 22:$\underline{5}$, $\overline{23}$:$\overline{25}$, $\underline{12}$:$\underline{26}$, $\underline{13}$:33, $\underline{7}$:27, 19:$\underline{13}$,

$\overline{8}$:$\overline{6}$, 14:$\underline{16}$, 35:$\underline{8}$, 23:$\underline{8}$, 11:$\overline{16}$, 35:$\overline{12}$

4. All effects, LASSO (109 variables, first 50 shown):

$\overline{24}$:$\overline{26}$, $\overline{20}$:$\overline{26}$, $\overline{18}$:$\overline{25}$, $\overline{7}$:$\overline{24}$, 6:$\overline{24}$, $\overline{22}$:$\overline{26}$, 1:$\overline{30}$, 1:$\overline{25}$, 18:$\overline{25}$,

4:$\overline{23}$, 1:$\overline{32}$, 32:$\underline{7}$, 8:$\overline{32}$, 23:$\underline{4}$, 13:$\overline{31}$, 3:$\overline{3}$,

31:$\underline{26}$, 2:$\overline{30}$, $\overline{22}$:$\overline{36}$, 31:$\underline{18}$, $\overline{10}$:$\overline{25}$, $\overline{24}$:$\overline{27}$, 26:$\overline{22}$,

26:$\overline{20}$, 17:$\overline{22}$, 20:$\overline{36}$, $\overline{10}$:$\overline{32}$, 31:$\underline{10}$, $\overline{16}$:$\overline{31}$, $\overline{27}$:$\overline{31}$

30:$\underline{10}$, 23:$\underline{9}$, 4:$\overline{36}$, $\overline{5}$:$\overline{27}$, $\overline{6}$:$\overline{18}$, $\overline{9}$:$\overline{32}$, 7:$\overline{18}$,

$\overline{4}$:$\overline{36}$, $\overline{11}$:$\overline{20}$, 4:7, 13:$\overline{24}$, 21:$\underline{27}$, $\overline{31}$:$\overline{32}$, 31:$\overline{31}$,

31:$\underline{4}$, $\overline{27}$:33, 3:$\underline{9}$, 16:$\overline{31}$, 26:$\overline{27}$, $\overline{20}$:$\overline{23}$

Table 3. Chosen variables for the best model in each category. Variables are listed in descending order of coefficient magnitude. See the text for a description of the notation.

A compact notation is used in the table to reduce the space consumed by long lists of variables. In this notation, each of the 35 spectral bands in the original images (the main effects) is represented by its band number. Squared terms are written with a bar over the band number, and square root terms are written with a bar underneath. Interactions between two terms are indicated by a colon. So, for example, the notation $\underline{9}$ refers to the square root of band 9, and 11:$\overline{17}$ refers to the interaction between band 11 and the square of band 17.

4.2. Predictive performance

The final estimate of the performance of the four selected models is based on those models' predictions on the complete set 37 test images. Together these images contain over 43 million pixels that were not used in any way during the model fitting and variable selection processes.

Because they are previously unused, they provide a more accurate approximation of the predictive power of the models (better than the validation data, which was not used for parameter estimation, but was used repeatedly for variable selection). The results are shown in Table 4.

	OER (%)	CER0 (%)	CER1 (%)
Model 1: RGB image	10.4	0.5	98.6
Model 2: main effects, 20 variables	8.6	0.5	82.1
Model 3: all effects, 50 variables	8.1	1.9	63.5
Model 4: all effects, LASSO (109 variables)	7.8	1.2	66.0

Table 4. Summary of the selected models and their predictive performance on the test images.

Figure 4 illustrates the trade-off between the different error types as the cutoff c is varied, for the 50-variable all effects model. The plot shows OER, CER0, and CER1 as functions of the cutoff. We can see that the overall error rate is in fact minimized at the original cutoff of 0.5, so changing the cutoff to improve performance on the smoke class will unfortunately come at the cost of worse overall performance. This notwithstanding, both OER and CER0 are relatively flat over the cutoff range (0.3, 0.5). So, for example, setting the cutoff to 0.4 will reduce the classwise error rate of smoke pixels to 50%, while increasing the OER only slightly.

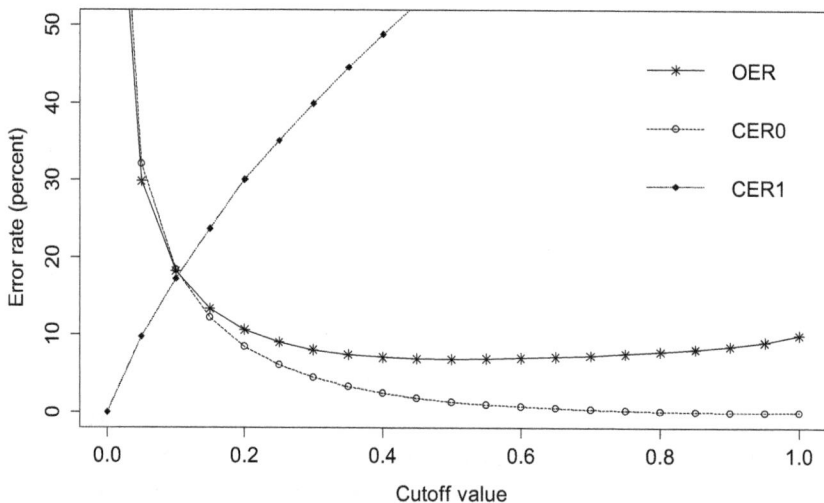

Figure 4. The effect of the decision cutoff on the overall and classwise error rates, for model 3.

5. Discussion

The experimental results are interpreted and discussed below, beginning with several remarks about model selection and performance evaluation, and followed by a qualitative evaluation of the classification results. Afterwards, a variety of suggestions for further improvement are provided.

5.1. Remarks on the selected models

The classification error rates were reported in Table 2 (for all models, on the validation set) and Table 4 (for the best models in each group, on the test set). Considering these tables, we see that our concern about the dominance of the smoke class (class 0) in the data set was justified. All of the models had overall error rates less than about 10%, which seems good at first glance. However in all cases this low error rate was achieved by having a very low error rate in the nonsmoke class (CER0) and a high error rate in the smoke class (CER1). This problem is particularly severe for smaller models and smaller sets of candidate variables, but even the best model in group 3 (the 50-variable model) had 56% misclassification of the smoke pixels.

Comparing the best models from each group, the only two models that can be considered even moderately successful are the two largest ones, the 50-variable all effects model (model 3) and the 109 variable LASSO model (model 4). There is little to separate these two classifiers: both have overall error rates of about 8% on the test set, with model 4 having a slight advantage; but model 3 has better performance on the smoke class.

Interestingly, these two models share only one variable in common (it happens to be 11:6). This is a consequence of the huge feature space and of the correlations among predictors. Two different models containing disjoint sets of variables can both have similar predictive power. This observation is related to the following two remarks.

Remark 1: physical interpretability of selected variables. It is desirable from a scientific and intellectual standpoint to be able to interpret the structure of a predictive model in terms of physical principles, but this is not always straightforward in a machine learning context. In the case of the spectral signature of smoke, a few general characteristics have been observed. Smoke scatters visible light [20], a component of it (organic carbon) is strongly absorbing below about 0.6 μm [21], and it is largely transparent in the middle infrared [22, 23]. We endeavored to interpret our models in light of these observations, but were unable to find any simple and unambiguous relationships based on the patterns of variables included in the models. This is often the price to pay for focusing on out-of-sample predictive accuracy: the classifier becomes a "black box" with internal structure that defies simple interpretation.

Remark 2: interpretability of model coefficients. Noticeably absent from the discussion so far has been the actual values of the regression coefficients in the fitted models. This has been deliberate, because in a pure classification problem like this one the predictive performance of the model as a whole is the overriding concern. Interpretability of model coefficients is desirable, but is likely not achievable when we have models with dozens of predictors that are all interactions. Assessment of statistical significance of particular predictors also adds nothing to our understanding of the model as a classifier, and is best avoided.

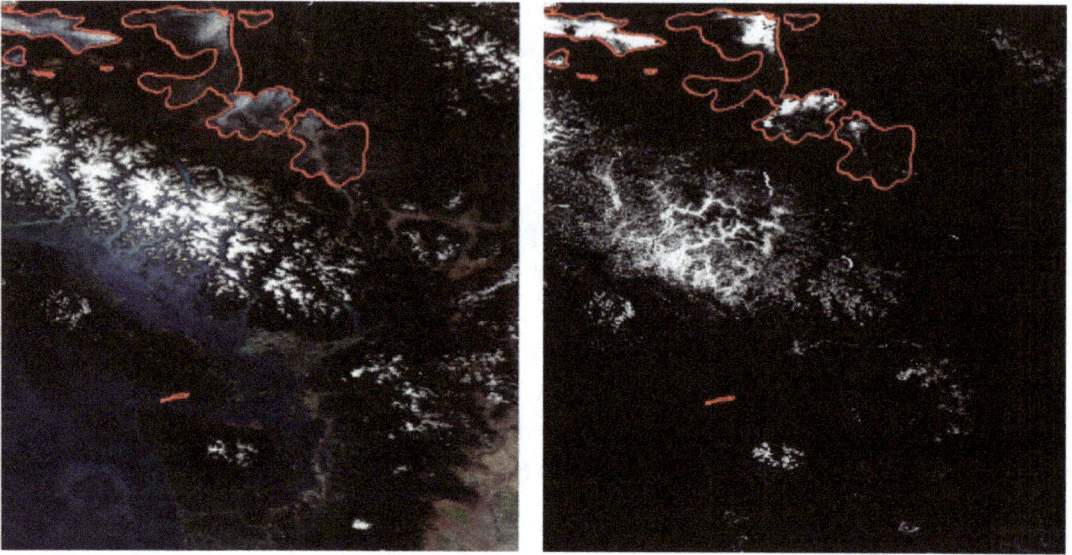

Figure 5. Results on a test image. Left: the RGB image. Right: the predicted probability map using the 50-variable model. The red contour delineates the true smoke region.

5.2. Qualitative performance analysis

Based purely on the observed numerical measures of prediction accuracy, it seems clear that none of the classifiers considered have performance good enough for real-world application, primarily because the majority of smoke pixels are misclassified in all cases. Visual inspection of the predictions on the test images can yield further insight into the nature of the problem, and possible causes of difficulty. Figure 5 and Figure 6 provide prototypical examples drawn from the test images. Our qualitative conclusions about predictive performance, based on the full set of 37 images, are listed below.

1. *Smoke-free images are generally classified well.* The classifier does have *some* ability to detect smoke, so it is still encouraging to observe that smoke-free images, or large regions that are smoke-free, are generally classified accurately. This can be observed in the bottom and left portions of Figure 6, which are assigned low probabilities throughout, despite the presence of clouds, water, and various types of terrain.

2. *Clouds and smoke can be distinguished well from one another.* It was observed that throughout the 37 test images, there were very few instances where cloud was erroneously identified as smoke. This provides at least some encouragement that the use of hyperspectral data holds benefits, because distinguishing clouds from smoke visually using the RGB images can be quite difficult.

3. *Snow and ice can be distinguished from smoke, but with greater difficulty.* A similar comment can be made about snow and ice, but less emphatically. The classifier generally performed well in separating smoke from snow and ice, but performance was less consistent. In certain images this task seemed to pose no problem, while in other images significant

numbers of snow or ice pixels were incorrectly labelled smoke. Both Figure 5 and Figure 6 provide some evidence of this, with moderate probabilities being mapped over the Coast Mountains in the upper left of either image.

4. *Co-located smoke and clouds present a problem.* The starting point for this problem is the assumption that smoke and clouds may both exist in the same pixel. Separation of smoke from clouds when both are in the same vicinity is a problem in two respects. First, when the masks were being prepared it was extremely difficult for the human interpreter to decide whether or not a given pixel in a cloudy region actually contains smoke. When clouds and smoke are mixed or adjacent, it is very difficult to distinguish one from the other using the RGB image alone. Second, because cloud is a significant constituent of the nonsmoke pixel class, the classifiers learned to assign low probability to pixels with the characteristics of clouds. An example of this problem can be seen in the upper right corner of Figure 6. In the RGB image, it is unclear if the bright feature in this corner is a cloud, and if so, whether there is also smoke present. From the probability map, it appears that there was indeed cloud in this region, which caused it to be assigned low probability.

5. *Prediction maps are unrealistically noisy.* Our mental model of the true scene in these images is of smoke regions being contiguous with relatively smooth boundaries. Because we are classifying pixels independently, however, this information is not incorporated into our procedures. The noisy nature of the probability maps is visible in both the smoke and nonsmoke regions ofFigure 5 andFigure 6.

Figure 6. Another example prediction. Top: RGB image. Bottom: predicted probability map.

6. *The quality of the training data is a major impediment to classifier construction.* Perhaps the most significant problem inherent in this study is uncertainty about the assigned classes in the original images themselves. Various portions of the images proved extremely difficult to assign to one class or the other with high confidence during the masking step. The aforementioned regions of mixed smoke and cloud provide one example. Regions where smoke becomes less concentrated provide another example (see Figure 5): where does the smoke end and the nonsmoke begin? In the same figure, we see a third example. A large number of pixels in a region over the mountains are "erroneously" assigned a high probability of being smoke. Is this a classification error, or an error in masking the original RGB image? The RGB image has a hazy appearance in this region, but it was not assigned to the smoke class due to the absence of a local fire and the general uncertainty about the nature of this hazy appearance. After the fact, it seems plausible that the classifier is detecting smoke that was erroneously labelled nonsmoke in the data set.

5.3. Opportunities for improvement

While the classification results were mixed, we feel there were enough positive elements to warrant further investigation, and that the overall approach can still be successful with appropriate modifications and extensions.

Probably the clearest opportunity for improvement is to alleviate the uncertainty in the true class labels that exists throughout the data set, and was illustrated in Figure 5 and Figure 6. The ambiguity in distinguishing smoke from nonsmoke at various places in the RGB images is a fundamental limitation. Simple approaches to solving this problem include considering only smoke plumes or "thick" smoke; excluding pixels that the photointerpreter finds ambiguous or that contain both cloud and smoke; or labelling images with more than two classes. More involved approaches include modelling each pixel as a mixture of different components, or modelling some continuous measure of smoke concentration rather than a binary presence/absence response. An unsupervised learning (clustering) approach or a semi-supervised method (where only some pixels are labelled) could also be considered, though such methods make quantitative performance assessment more difficult.

Another avenue for potential improvement of classification performance is to modify the feature space in the logistic model in the hopes of improving the separability of smoke and nonsmoke. While this could be done by adding even more factorial terms (cubic terms, higher-order interactions, and so on), it is unlikely that the benefit of doing so would outweigh the increase in computational burden. Instead, more focused modifications of the model could be considered. To reduce the effect of highly heterogeneous surface terrain in the nonsmoke class, for instance, a baseline spectrum (perhaps taken as an average of observations over recent clear-sky days) could be included as predictors in the model. Or each pixel could be assigned to a known ground-cover class at the outset, and these classes could be included in the model as categorical variables. Another option is to replace the fixed powers of reflectance we used (squared and square root terms) with spline functions, allowing data-adaptive nonlinear transformations of the variables to be used in the model. We anticipate exploring some of these alternatives in future work with these data.

Additional possibilities for improvement can be found by moving farther from the logistic regression framework. Under the assumption of independent pixels, for example, any of the many existing classification tools could be applied to the data. The support vector machine (see, e.g., [24], Ch. 11) in particular is a state-of-the-art method that has performed well across a variety of tasks and is worthy of consideration. If the independence assumption is dropped, the autologistic regression model [25], a model for spatially-correlated binary responses, is a natural fit for these data. This model would alleviate the problem of noise in the predicted probabilities, producing smoother and more accurate prediction maps. It is a natural extension of logistic regression to spatially-associated data. Finally, it may also be possible to incorporate relevant ancillary information (for example, prior knowledge of fire locations and wind directions) into a classification model to improve predictive power. Again, consideration of these alternatives and extensions are planned in future work.

6. Conclusion

The smoke identification problem provided a case study on the use of supervised learning to automate the process of recognizing features of interest in remote sensing images. The machine learning approach is especially attractive when working with hyperspectral images, because the high dimensionality of the data makes it very challenging for a human photointerpreter to consider all of the potential relationships in the data. Subject-matter knowledge can help to focus a human expert on certain models, relationships, or spectral bands, but automated procedures provide a valuable complementary approach. They can be used to search for more complex or previously unconsidered relationships, driven by the data itself. If a machine learning procedure can be implemented successfully, another clear benefit is the ability to process data at a speed and scope not feasible by other means.

Our primary conclusion regarding the smoke identification goal is that the spectral information in the smoke and nonsmoke classes overlap to such a degree that it is not possible to construct a highly successful classifier—at least with the models and methods we employed. The results have some promising elements, however. Notably, it appears possible to distinguish smoke from cloud and snow when a) the smoke is not mixed with cloud, and b) the smoke is not too diffuse. Indeed, if the goal of the study were to find clear-sky smoke plumes only, the approach would be quite successful. Classification errors were largely attributable to the presence of cloud in a smoky region, to the smoke being too diffuse, or to inaccuracies introduced in the initial labelling of the data. Armed with this understanding, it should be possible to make considerable improvements to the results with adjustments to the methodology.

The problem used for this case study is a challenging image segmentation task, made more challenging by the loose definition of "smoke" used in the initial labelling of the data set. Reflecting this, the best classifiers we found were only partially successful. Still, the process of developing them has helped to provide insight into the problem and allows us to present both the advantages and challenges of the machine learning approach. With the dimensionality and throughput of remote sensing data ever on the rise, computer intensive techniques such as those explored here will be of increasing importance in the future.

Author details

Mark A. Wolters[1*] and C.B. Dean[2]

*Address all correspondence to: mwolters@fudan.edu.cn

1 Shanghai Center for Mathematical Sciences, Fudan University, Shanghai, China

2 Department of Statistical and Actuarial Sciences, Western University, London, Canada

References

[1]   Richards JA. Remote Sensing Digital Image Analysis. 5th ed.: Springer-Verlag; 2013.

[2]   R Core Team. R: A Language and Environment for Statistical Computing Vienna: R Foundation for Statistical Computing; 2014.

[3]   National Aeronautics and Space Administration. MODIS Web. [Online].; 2014 [cited 2014 November 25. Available from: http://modis.gsfc.nasa.gov/about/specifications.php.

[4]   MODIS Characterization and Support Team. MODIS Level 1B Product User's Guide. Greenbelt, MD: NASA/Goddard Space Flight Center; 2012. Report No.: MCST Document # PUB-01-U-0202-REV D.

[5]   Goddard Space Flight Center. LAADS Web. [Online].; 2014 [cited 2014 November 25. Available from: http://ladsweb.nascom.nasa.gov/.

[6]   Gumley L, Descloitres J, Schmaltz J. Creating Reprojected True Color MODIS Images: A Tutorial. Maison, WI: University of Wisconsin-Madison, Space Science and Engineering Center; 2010. Report No.: Version 1.0.2.

[7]   Giglio L, Descloitres J, Justice CO, Kaufman YJ. An Enhanced Contextual Fire Detection Algorithm for MODIS. Remote Sensing of Environment. 2003; 87: p. 273-282.

[8]   Bishop CM. Pattern recognition and machine learning New York: Springer; 2006.

[9]   Hastie T, Tibshirani R, Friedman J. The Elements of Statistical Learning: Data Mining, Inference, and Prediction. 2nd ed.: Springer Science+Business Media; 2009.

[10] James G, Witten D, Hastie T, Tibshirani R. An Introduction to Statistical Learning with Applications in R: Springer Science+Business Media; 2013.

[11] Tibshirani R. Regression Shrinkage and Selection via the Lasso. Journal of the Royal Statistical Society, Series B. 1996; 58(1): p. 267-288.

[12] Efron B, Hastie T, Johnstone I, Tibshirani R. Least Angle Regression. The Annals of Statistics. 2004; 32(2): p. 407-451.

[13] Friedman J, Hastie T, Tibshirani R. Regularization Paths for Generalized Linear Models via Coordinate Descent. Journal of Statistical Software. 2010; 33(1).

[14] Miller A. Subset Selection in Regression. 2nd ed.: Chapman & Hall; 2002.

[15] Wolters MA. A Genetic Algorithm for Selection of Fixed-Size Subsets, with Application to Design Problems. Journal of Statistical Software. (to appear).

[16] Michalewicz Z, Fogel DB. How to Solve it: Modern Heuristics. 2nd ed.: Springer-Verlag; 2004.

[17] Rothlauf F. Design of Modern Heuristics: Principles and Application Heidelberg: Springer; 2011.

[18] Gendreau M, Potvin JY, editors. Handbook of Metaheuristics. 2nd ed.: Springer; 2010.

[19] Whitley D, Beveridge JR, salcedo CG, Graves C. Messy Genetic Algorithms for Subset Feature Selection. 1997..

[20] Li Y, Vodacek A, Zhu Y. An automatic statistical segmentation algorithm for extraction of fire and smoke regions. Remote sensing of environment. 2007; 108(2): p. 171-178.

[21] Jethva H, Torres O. Satellite-Based Evidence of Wavelength-Dependent Aerosol Absorption in Biomass Burning Smoke Inferred from Ozone Monitoring Instrument. Atmospheric Chemistry and Physics. 2011; 11: p. 10541-10551.

[22] Miura T, Huete AR, van Leeuwen WJD, Didan K. Vegetation Detection Through Smoke-Filled AVIRIS Images: An Assessment Using MODIS Band Passes. Journal of Geophysical Research. 1998; 103(D24): p. 32001-32011.

[23] Chu DA, Kaufman YJ, Remer LA, Holben BN. Remote Sensing of Smoke from MODIS Airborne Simulator During the SCAR-B Experiment. Journal of Geophysical Research. 1998; 103(D24): p. 31979-31987.

[24] Izenman AJ. Modern Multivariate Statistical Techniques: Springer Science+Business Media; 2008.

[25] Hughes J, Haran M, Caragea PC. Autologistic models for binary data on a lattice. Environmetrics. 2011; 22(7): p. 857-871.

## 10

Sea Transport Air Pollution

Ivan Komar and Branko Lalić

Additional information is available at the end of the chapter

1. Introduction

The aim of this chapter is to provide an overview of the air pollution generated by diesel engines of the ocean-going ships and the technologies as well as methodologies available to reduce these emissions. This chapter begins with general significant information of the air pollutant emission from ships followed by a summary of the International Maritime Organization (IMO) regulatory MARPOL Annex VI being developed to control marine shipping emissions as well as information on the various types of the ocean-going ships and their prime movers with particular emphasis on marine diesel engines as sources of air pollution from ships. For better understanding of the formation of air pollutants from marine diesel engines, authors gave a brief overview of the working principles of marine diesel engines as well as their combustion process and chemistry of the pollutant formation during that process. Finally, the chapter concludes with an analysis of several control methods that can effectively reduce harmful pollutant emissions from marine diesel engines.

Climate change on Earth is one of the largest civilised problems at the beginning of the twenty-first century. Anthropogenic impact on the Earth's climate became one of the crucial environmental issues of modern civilisation in the late twentieth century. Therefore, nowadays the ecology and preservation of human environment have become two of the very important human activities all over the world. Besides primary pollution from the land, nowadays attention is being paid to the pollution from the ships. In recent decades, shipping industry and maritime traffic have rapidly developed. From the economic point of view, this trend, which continues today, has a very positive impact on economic development but on the other hand, a very negative impact on the environment in terms of air pollution. Exhaust gases from marine diesel engines are the primary source of emissions from ships and contribute significantly to environmental pollution. Ocean-going ships are the major contributors to global emissions of several hazardous air pollutants such as nitrogen oxides (NOx), sulphur oxide

(SOx), fine particulate matter (PM), hydrocarbons (HCs), carbon monoxide (CO) and green-house gas carbon dioxide (CO_2).

The presence of these pollutants has local and global impact. Impacts on local (or regional) air quality are mainly linked to pollutants such as PM, NOx and sulphur, while CO_2 has a global impact on climate [1]. The amount of gases emitted from marine engines into the atmosphere is directly related to the total fuel oil consumption. While pollutant emissions from land-based sources are gradually decreasing, those from shipping show a continuous increase. It is estimated that by 2020, the emissions NOx and SOx from international shipping around Europe are expected to equal or even surpass the total emissions from all land-based mobile, stationary and other sources in the 25 EU member states combined (see Figure 1). It should be noted that these figures refer only to ships in the international trade and do not include emissions generated from shipping in countries' internal waterways or from ships plying harbours in the same country, which are given in the domestic statistics of each country [2].

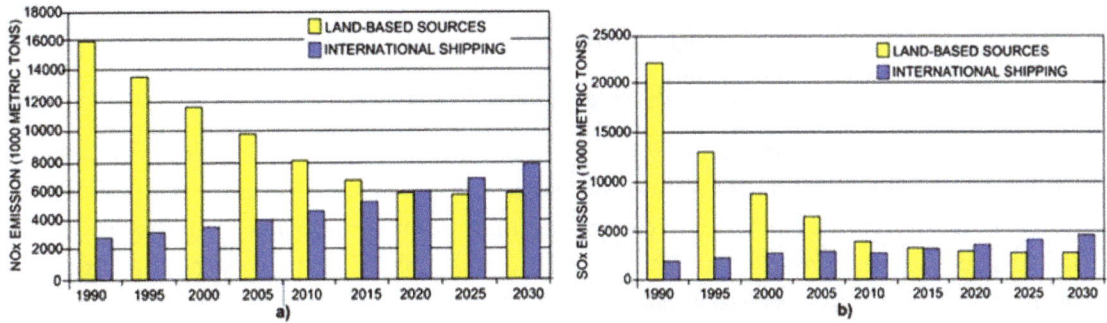

Figure 1. Emissions NOx (a) and SOx (b) 1990–2030 from land-based sources and international shipping

It has been estimated that about 90 % of the total sulphur dioxide (SO_2) and NOx emissions from ships in the North Sea, including the English Channel, originate from a zone of approximately 50 nautical miles from the coast line. International shipping was estimated to be a source of 97 % of the total SO_2 and NOx emissions in the North Sea within a distance of 100 nautical miles from the seaboard, as in [3]. The emissions of sulphur dioxide (SO_2), particulate matter (PM) and greenhouse gases (GHGs) from global shipping were increased from 585 to 1096 million tons between 1990 and 2007. The CO_2 emissions from international shipping are estimated at 943.5 million tons for the year 2007 and CO_2 emissions from global shipping are about 1 billion tons for the year 2006. International shipping is responsible for 3 % of global CO_2 emissions as in [4].

2. International regulation concerning air pollution from merchant shipping

Regulations concerning air pollution from merchant shipping are developed at the global level. Because shipping is inherently international, it is vital that shipping is subjected to uniform regulations on issues such as air emissions from ships. The shipping industry is principally regulated by the International Maritime Organization (IMO), which is a UN agency based in London and responsible for the safety of life at sea and the protection of the marine environment. IMO ship pollution rules are contained in the 'International Convention on the Prevention of Pollution from Ships' known as MARPOL 73/78, which represents the first set of regulations on marine exhaust emissions. The original MARPOL Convention was signed on 17 February 1973, but did not come into force. The current Convention is a combination of the 1973 Convention and the 1978 Protocol. MARPOL73/78 contains 6 annexes concerned with preventing different forms of marine pollution from ships:

- Annex I deals with regulations for the prevention of pollution by oil.

- Annex II details the discharge criteria and measures for the control of pollution by noxious liquid substances carried in bulk.

- Annex III contains general requirements for issuing standards on packing, marking and labelling.

- Annex IV contains requirements to control pollution of the sea by sewage.

- Annex V deals with different types of garbage, including plastics, and specifies the distances.

- Annex VI deals with gaseous emissions of ship engines and installations: the Convention regulates sulphur oxide, nitrogen oxide and particulate matter emissions from ship exhausts and prohibits deliberate emissions of ozone-depleting substances. It also contains provisions allowing for the creation of special Emission Control Areas (ECA) with even more stringent controls on air pollutant emissions. Annex VI also forbids any (deliberate) emission of an ozone-depleting substance, such as halons and chlorofluorocarbons (CFCs) as well as any new installation of equipment using these gases. Annex VI entered into force on 19 May 2005 and sets limits on nitrogen oxides, sulphur oxides and volatile organic compound (VOC) emissions from ship exhausts and prohibits deliberate emissions of ozone-depleting substances [5]. Annex VI regulation 13, nitrogen oxides (NOx), applies to diesel engines over 130 kW installed on ships built on or after 1 January 2000, excluding engines for emergency purposes such as emergency generator engine, lifeboat engine, etc.

Three different levels (tiers) of NOx control apply based on the ship construction date as follows:

- *Tier 1* entered into force in 2005 and applies to marine diesel engines installed in ships constructed on or after 1 January 2000 and prior to 1 January 2011.

- *Tier 2* entered into force in 1 January 2011 and replaced the Tier 1 NOx emission standard globally. It applies globally for new marine diesel engines installed in ships constructed on 1 January 2011 or later. Tier 2 NOx emission levels correspond to about 20 % reduction from the Tier 1 NOx emission standard. Tier 2 is applicable outside the Tier 3 designated Emission Control Areas (ECA).[1]

- *Tier 3* will enter into force in the year 2016 and it will by then apply for new marine diesel engines > 130 kW installed in ships constructed on 1 January 2016 or later when operating inside the ECA. The Tier 3 NOx emission level corresponds to an 80 % reduction from the Tier 1 standard. The NOx emission limits are expressed as dependent on engine speed (n) in revolution per minute (RPM). These are shown in Table 1 and Figure 2 [6,7].

Tier	Effective date	NOx limit (g/kWh)		
		$n < 130$	$130 < n < 2000$	$n > 2000$
Tier I	2000	17	$45\,n^{-0,2}$	9,8
Tier II	2011	14,4	$44\,n^{-0,2}$	7.7
Tier III	2016	3,4	$9\,n^{-0,2}$	1,96

Table 1. NO_x limits according to MARPOL Annex VI

For engines with an engine speed lower than 130 RPM, the Tier III level is 3.4 g/kWh. When operating outside an ECA, the engine must meet the Tier II limit of 14.4 g/kWh. Engines with an engine speed higher than 130 RPM must meet even lower limits (see Table 1 and Figure 2). Any abatement technology reducing the NOx emission to the required level can be accepted.

Furthermore, MARPOL Annex VI has set a maximum global fuel sulphur limit of currently 3,5 % in weight (from 1 January 2012) for any fuel used on board a ship. Annex VI also contains provisions allowing for special SOx Emission Control Areas (ECA) to be established with more stringent controls on sulphur emissions. In an ECA, the sulphur content of fuel oil used on board a ship must currently not exceed 1 % in weight. The MARPOL Annex VI has undertaken a review with the intention to further reduce emissions from ships. The current and upcoming limits for future fuel oil sulphur contents are presented in Table 2 and Figure 2 [6,7].

Fuel sulphur cap	Area	Date of implementation
Max. 1 % S in fuel	ECA	1 July 2010
Max. 3,5 % S in fuel	Globally outside ECA	1 January 2012
Max. 0,1 % S in fuel	ECA	1 January 2015
Max. 0,5 % S in fuel	Globally outside ECA	1 January 2020

Table 2. Sulphur limits in fuel according to MARPOL Annex VI

The rules of SOx apply to all ships, no matter the date of ship construction. Although the SOx requirements can be met by using a low-sulphur fuel, the regulation allows alternative methods to reduce the emissions of SOx to an equivalent level.

---

1 Designated Emission Control Areas (ECA) that are defined by the IMO currently comprise the Baltic Sea, the North Sea, the English Channel, the US Caribbean Sea and the area outside North America (200 nautical miles – see Figure 3) The first sulphur Emission Control Area (ECA) was established in the Baltic and came into force internationally on 19 May 2005, and all ships were required to either use the 1.5 % low-sulphur fuel or fit an exhaust gas cleaning system as required by regulation 14 of Annex VI 12 months from this date which was 19 May 2006.

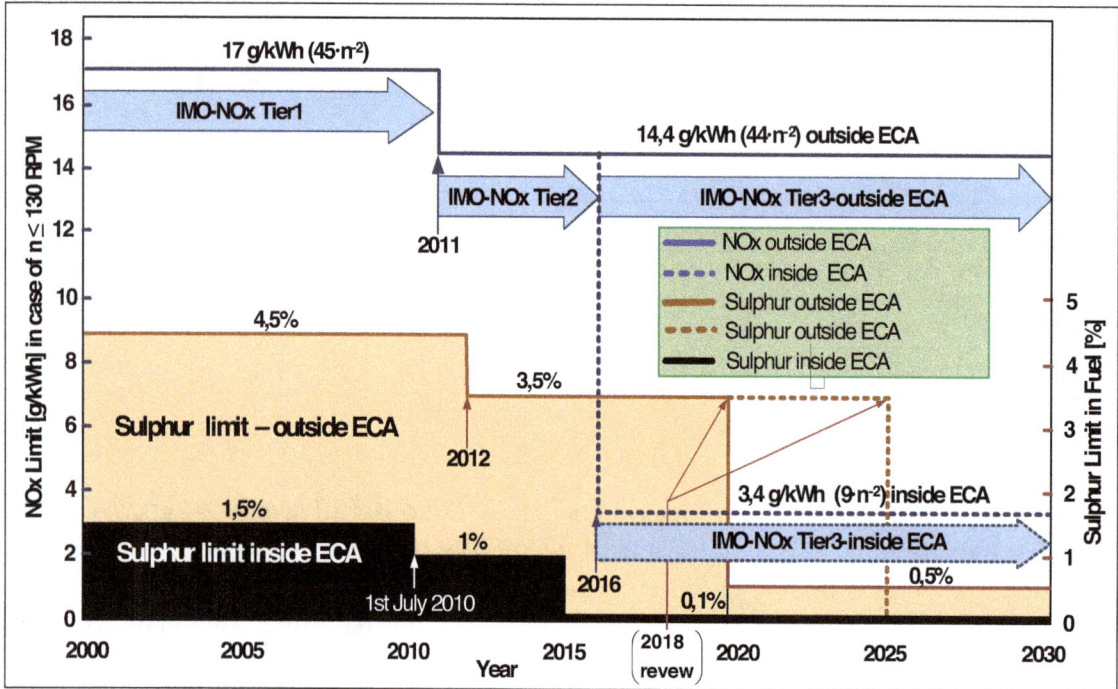

Figure 2. Reduction of NOx and sulphur in fuel on the global and ECA limit

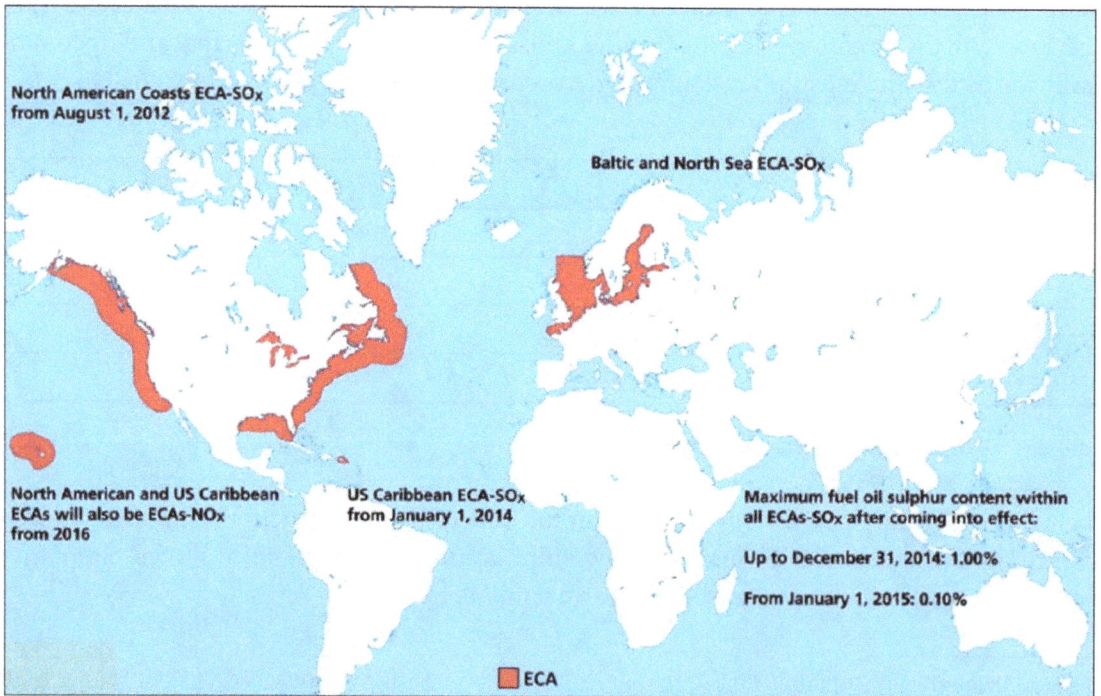

Figure 3. Existing Emission Control Areas as per MARPOL Annex VI (source: www.marlink.com)

The sulphur ECA limit which entered into force in 1 January 2015 can be met using a low-sulphur fuel with sulphur below 0.1 %. The global limit, outside sulphur ECA, can be met using fuel with sulphur content below 0.5 %, which will be required from 2020. The date of the global limit reduction may be changed to 2025 as a result of a feasibility review to be conducted no later than 2018 (see Figure 2).

In order to comply with the requirements of Regulation 14 of MARPOL, the burning of low-sulphur fuel oils (LSFOs) was introduced. There is also an EU regulative about the sulphur content in marine gas oil. Namely, in accordance with EU's marine fuel sulphur directive (1999/32/EG, Article 4 with amendment as per directive 2005/33/EC), the sulphur content in marine gas oil within the territorial waters of an EU member state may not exceed 0.1 % by weight. This applies to all ships regardless of flag. As of 1 January 2010, the sulphur content of any marine fuels may never exceed 0.1 % by weight for ships in port with the exception of short stays in port (up to 2 h).

For ships continuously operating on low-sulphur fuel oil and for those that may be fitted with an exhaust cleaning system, there is no issue. However, for ships which burn heavy fuel oil with 3.5 % sulphur content and are not equipped with scrubbing equipment, the problem of compliance is much greater as large quantities of fuel are being mixed during the oil change-over to low-sulphur fuel oil. MARPOL Annex VI Regulation 14 requires those ships using separate low-sulphur fuel to comply with this regulation and in entering or leaving an Emission Control Area, shall carry a written procedure showing how the fuel oil changeover is to be done, allowing sufficient time for the fuel oil service system to fully flushed out all fuel oils exceeding the applicable sulphur content prior to entry into a sulphur Emission Control Area. The volume of low-sulphur fuel oils in each tank as well as the date, time and position of the ship when any fuel oil changeover operation is completed prior to the entry into an ECA or commenced after exit from such an area shall be recorded in such logbook as prescribed by the Annex VI Regulation.

Furthermore, ships are also an important source of greenhouse gas (GHG) pollutants. According to the Green House Gas study by the IMO consensus, international shipping emitted 843 million metric tonnes of carbon dioxide, 2.7 % of the global CO_2 emissions in 2007. Including domestic shipping and fishing ships larger than 100 gross tonnes (GT),[2] the amount would increase to 1.019 billion tonnes, i.e. 3.3 % of the global CO_2 emissions. At the present trend, this percentage could go two or three times higher from the present by 2050 emissions [8, 9].

In order to control this CO_2 emission from shipping, the first formal CO_2 control regulations were adopted by the IMO at the 62nd session of the Marine Environment Protection Committee (MEPC) in July 2011. The amendments to MARPOL Annex VI included the addition of Chapter 4 on regulations on energy efficiency for ships to make mandatory the Energy Efficiency Design Index (EEDI) for new ships and the Ship Energy Efficiency Management Plan (SEEMP)

---

2 Gross tonnage (GT) determined under the IMO 1969 Tonnage Convention represent the total volume of of ship's enclosed spaces measured in register tonne (RT) where 1RT=100 f3 or 2,83 m3, conversely. Net tonnage (NT) determined under the IMO 1969 Tonnage Convention represents the volume of cargo spaces (cargo holds and cabins) in register tonne (RT).

for all ships. The regulations apply to all ships of 400 GT and above and are entered into force in 1 January 2013.

The basic formulation of EEDI is based on the ratio of total CO_2 emission per tonne mile. As CO_2 depends upon fuel consumption and fuel consumption depends upon the total power requirements, eventually this EEDI formulation has certain impact on ship design parameters and hydrodynamics. The SEEMP establishes a mechanism for a shipping company and a ship to improve the energy efficiency of ship operations. The SEEMP provides an approach for monitoring ship and fleet efficiency performance over time using. The results from the study that IMO ordered from Lloyd's Register and Det Norske Veritas to estimate the impact of the new requirements show that the EEDI will, as new ships are built, gradually reduce the emissions from the world fleet with 3 % in 2020, 13 % in 2030 and 30 % in 2050. The SEEMP will not directly mandate an emission reduction, but by increased awareness of costs and reduction potentials, the study estimated the reduction to be between 5 and 10 % from 2015 onwards [9,10].

3. Sea ship classification and quantification

As per rules of the Shipping Classification Societies, ship is defined as 'a floating unit intended for sea-going service with length greater than 12 meters and with GT greater than 15, or which carries more than 12 passengers. The present definition does not apply to ships of war and troopships'. Marine ocean-going ships are generally very large ships designed for deepwater navigation. Depending on the nature of their cargo, ships can be divided into different categories, classes and types. A majority of these ships can be classified as one of the following: tanker, bulk carrier, container ship, ro-ro ship, general cargo ship, reefer ship and passenger ship. There are also smaller ship types, which are not included in the largest categories of ship, as fishing ships intended and equipped for fishing or exploiting other living resources of the sea; tugs, a ship specially constructed and equipped for towing and/or rescuing and salvage of ships or other floating units; ships used by authorities which include the following types: pilot boats, rescue ships, police boats, custom boats, etc.; training ships provided for training of marine personnel gaining training and practical marine experience to develop seafaring skills suitable for a professional career at sea and provided with special equipment and arrangements suitable for that purpose (teaching rooms, accommodation spaces for teachers and trainees, etc.); research ship, a ship without cargo spaces, engaged in scientific research, noncommercial expeditions and surveys, carrying scientists, technicians and members of expeditions, and provided with special equipment and arrangements suitable for that purpose (i.e. laboratories, accommodation for research personnel, etc.); supply ship, a ship mainly intended and equipped for the carriage of special personnel, special materials and equipment which are used to provide facilities to offshore units and other marine installations, as well as to provide assistance in performing special activities; and icebreakers and recreational ships such as yachts classified as recreational craft for personal or commercial use, having hull length greater than 12 m, having facilities and accommodation for extended navigation, authorised to carry not more than 12 passengers, excluding crew. The following is a brief description of the characteristics of the main types of ocean-going ships:

Tanker is a merchant ship designed to transport liquids or gases in bulk. The major types of tanker ship include the oil tanker, the chemical tanker and gas carrier.

Oil tanker is a ship which is constructed primarily to carry oil in bulk and comes in two basic types: the crude carrier, which carries crude oil, and the clean product tanker, which carries the refined products, such as petrol, gasolene, aviation fuel, kerosene and paraffin. Tankers also include ship types such as combination carriers. Combination carrier is a general term applied to ships intended primarily to carry oil or dry cargoes, including ore, in bulk (ore/oil ships, oil/bulk/ore – OBO). These cargoes are not carried simultaneously. Generally they are constructed with a single deck, two longitudinal bulkheads and a double bottom throughout the cargo length area and intended primarily to carry ore cargoes in the centre holds or of oil cargoes in centre holds and wing tanks.

Chemical tankers are ships which are constructed generally with integral tanks and intended primarily to carry chemicals in bulk.

Gas carrier can be divided into two types: the *LNG tanker* carries liquified natural gas and the *LPG tanker* carries liquified petroleum gas. Tankers can range in size of capacity from several hundred deadweight tons (DWT),[3] which include ships for servicing small harbours and coastal settlements, to the real giants of several hundred thousand DWT: the VLCC (very large crude carrier) of between 200,000 and 300,000 DWT and the ULCC (ultra large crude carrier) of over 300,000 DWT.

Bulk carriers are sea-going self-propelled ships which are constructed and intended primarily to carry dry cargoes in bulk such as ore, coal, pulp, rock, cement, scrap metal, grain, flour, rice, fertilisers, sugar or any cargo that travels in bulk. Bulk carriers range from about 25,000 DWT ('handysize') through the medium-size ('Panamax') ships of about 75,000 DWT, to the giant ('capesize') ships of over 200,000 DWT.

Container ships are a type of dry cargo ships specially designed and equipped with the appropriate facilities for carriage of containers. They carry standardising container at 20-feet long (TEU – twenty-foot equivalent unit) or 40-feet long (FEU – forty-foot equivalent unit). Today's container ships are being built to take up to 18,000 TEU.

Ro-ro (roll on/roll off) is a cargo ship (ferry) specifically designed for the carriage of vehicles, which embark and disembark on their own wheels, and/or goods on pallets or in containers which can be loaded or unloaded by means of wheeled vehicles. Another type of ro-ro is a passenger ship (ROPAX). The acronym ROPAX (roll-on/roll-off passenger) describes a ro-ro ship built for freight vehicle transport along with passenger accommodation. Ro-ro ships have built-in ramps that allow the cargo to be efficiently rolled on and off the ship when in port. The ramps and doors may be stern-only, bow and stern or side for quick loading.

---

3 Deadweight tonnage (also known as deadweight abbreviated to DWT) is a measure of how much weight a ship is carrying or can safely carry. It is the sum of the weights in metric tonnes of cargo, fuel, freshwater, ballast water, provisions, passengers and crew.

General cargo ship is a ship intended for the carriage of general cargo which will not be carried in containers.

Refrigerated cargo ship or reefer ship is a ship (excluding liquefied gas carriers and fishing ships) specially intended to carry permanently refrigerated cargoes such as fruits, vegetables, dairy products, fish and meat and has fixed refrigeration installations and insulated holds. Excluding the temperature control, the reefers are similar to other dry cargo ships or containers.

Passenger ship as per rules of the ship's classification society is a self-propelled ship with a permission to carry more than 12 passengers, specially designed and equipped for that purpose, with a single or multi-deck hull and superstructure and with cabin accommodation for passengers.

As of 1 January 2013, there were 48,732 merchant ships (597,709,000 GT) involved in international trade, registered under the flags of over 150 nations (see Figure 4).

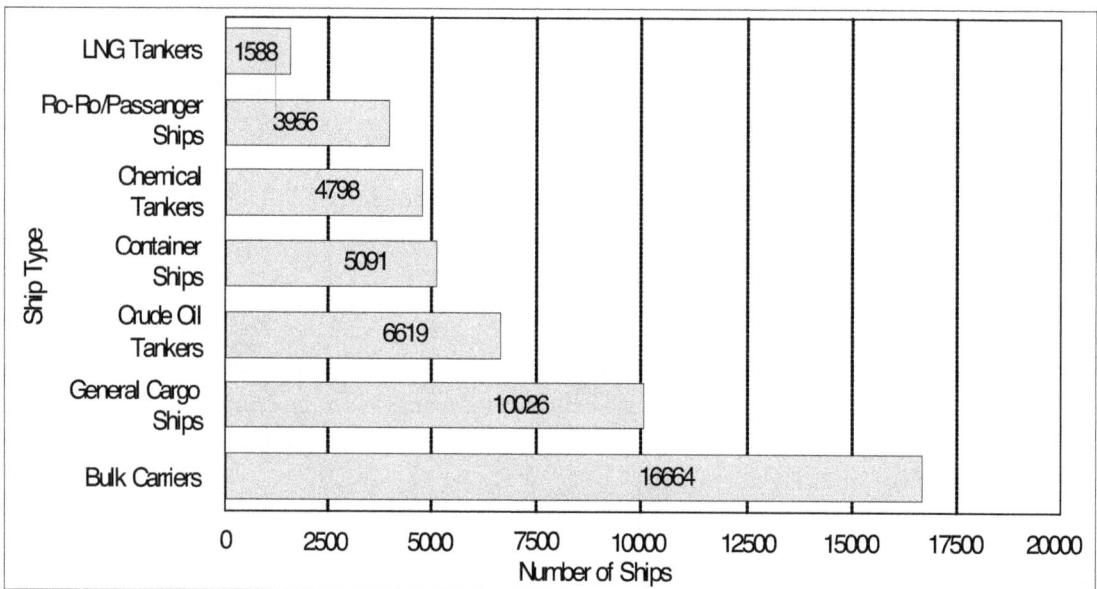

Figure 4. Number of the main types of ocean-going ships in the world merchant fleet as of 1 January 2013 (source: http://www.statista.com)

3.1. Ship engines

The power needed on ships is generated through main and auxiliary engines often called prime mover in the literature and can be sorted as diesel engine, gas turbine, steam turbine and electric motor. The diesel engine is the most common prime mover in the merchant marine, mainly due to its low fuel consumption in comparison with other prime movers. Power ranges between 0.25 MW for the smallest high-speed engines and 100 MW for the for the biggest low-speed marine diesel engines. The main advantages of diesel engines are the following: it is relatively insensitive to fuel quality; it can be operated by light fuel as well as the heaviest residual fuels; and it has high reliability and high efficiency. In the other hand, the main disadvantages of diesel engines are pollutant emissions, low power-to-weight ratio if compared with gas turbine and vibration and noise.

From the application viewpoint, three main types of diesel engines are available: low-speed diesel engines (rpm<200), medium-speed diesel engines (200<rpm<1000) and high-speed diesel engines rpm>1000).

From the construction viewpoint, two types can be distinguished: two-stroke low-speed engines and four-stroke engines (medium or high speed). Low-speed diesel engines are dominant in the deep sea tanker, bulk carrier and containership sectors. Such types of engines are used as ship's propulsion engine without gearbox, i.e. directly connected to the propeller shaft system. These engines are currently the most efficient in terms of the specific fuel consumption, but NOx emission level from these engines is very high in comparison with medium- or high-speed marine diesel engines (see Figure 5b). The diesel engines have specific fuel consumptions around 160–185 g/kWh, against the 220–240 g/kWh of gas turbines and 300 g/kWh of steam turbines (see Figure 5a).

Medium-speed engines are used for auxiliaries such as alternators and for the main propulsion engine, with a gearbox to provide a propeller speed reduction, for smaller cargo ships, ferries, passenger cruise liners, ro-ro carriers, supply ship, icebreakers, etc., while high-speed engines are used as propulsion and for auxiliaries in smaller ships as fishing ships, tugs, pilot ships, recreational ships, etc.

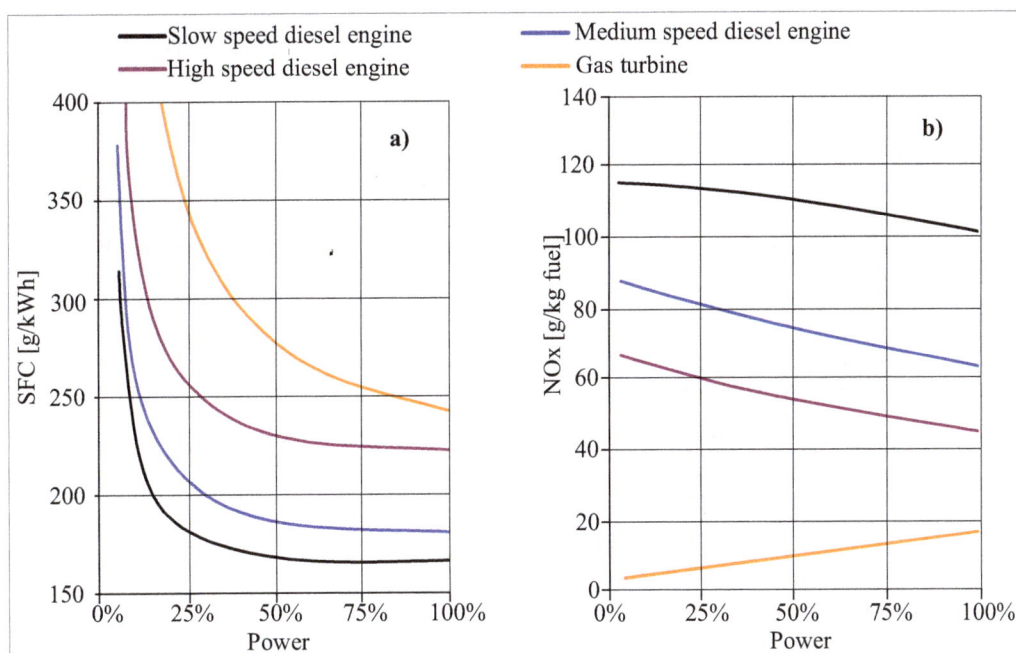

Figure 5. a) Specific fuel consumption of prime movers b) NOx emission ratio for prime movers

3.2. Marine diesel engine: Type and working principles

The diesel engine is reciprocating internal combustion engine where chemical energy of the fuel is converted into thermal energy by means of combustion reactions of the fuel, and then the thermal energy is converted into mechanical work. The actual cycle inside the engine can be done either in four strokes (two crank revolutions) or in two strokes (one crank revolution) [11]. A stroke is defined as the distance travelled by the piston between the top dead centre (TDC) and the bottom dead centre (BDC). Cycles in the four-stroke engine are compression, power, exhaust and intake strokes. During compression stroke, the piston moves upwards from BDC to TDC. Inlet and exhaust valves are closed, and the combustion air is compressed and thus increases air pressure and temperature (110 to 220 bar and 600 °C to 800 °C depending on the type of engine). Fuel is injected several crank degrees before TDC and ignited by the high temperature of the compressed air. Combustion starts at the end of the compression stroke. The combustion is continued over a considerable crank angle after TDC, while the combustion gases expand and perform work on the piston, forcing it down. This is power stroke. Towards the end of the stroke, the exhaust valve opens which releases the combustion gases into the exhaust manifold or exhaust gas receiver. Thereafter during the exhaust stroke, the piston moves from BDC to TDC. The exhaust valve is open and the rest of the combustion gases are forced out of the cylinder by the upward stroke of the piston. The gases that remain in the cylinder are dispelled by a scavenging process; the inlet valve is opened early, whereas the exhaust valve is closed late, so that both are open at the same time (overlap period). Thereafter during the intake stroke, the piston moves downwards from TDC to BDC. The inlet valve is open and the exhaust valve closed, while the cylinder is filled with a charge of fresh air and thereafter is ready for the compression stroke. A complete cycle takes four strokes. However, the power stroke is the only useful one, which suggests that there is a useful stroke every two crank revolutions.

Slow-speed marine diesel engine (65–200 rpm) operates on the two-stroke cycle. This means that this engine has one working or power stroke per every crank revolution. The main difference comparing two-stroke with the four-stroke cycle is that charging and exhaust take place without the piston enforcing the process. During the compression stroke, the inlet ports and exhaust valve are closed and a volume of air is trapped into the cylinder. The piston moves upwards to TDC, thus compressing this combustion air and causing a temperature and pressure rise (110 to 160 bar and 600 °C to 800 °C depending on the type of engine) that is sufficient to ignite the fuel that has been injected few degrees before TDC. At the end of the compression stroke, combustion starts and continues several degrees during power stroke. The combustion gases expand and perform work on the piston, forcing it down from TDC to BDC. At the end of expansion, exhaust valve opens and combustion gases blow down to exhaust receiver pressure. By the time the inlet ports are open, the cylinder pressure will have reached a pressure lower than that of the scavenging air, so scavenging starts. These processes take place in one stroke. Scavenging, which starts while the piston moves downwards, is completed while the piston moves upwards. Both the inlet ports and exhaust valve are opened, and fresh air from scavenging air receiver enters the cylinder, forcing the exhaust gases out. The scavenging and compression processes take place in another stroke. The process described here is for uniflow scavenging. It is necessary to compress the scavenging air by the turbo-charger in order to scavenge the cylinder. The principal scheme of a turbocharged marine diesel engine is shown in Figure 6 as in [12]. Turbocharger (TC) is composed of the compressor (C) driven by the gas turbine (T) that receives its power from the heat energy of the exhaust gases flowing through. The compressor and the turbine are directly coupled and they are built together in a common housing. The gas turbine, usually one-stage axial type, is located after an exhaust gas receiver (EGR) which collects the exhaust gases from all the engine cylinders. The one-stage centrifugal-type compressor feeds compressed air through scavenging air cooler into scavenging air receiver (SAR) that supplies all the engine cylinders.

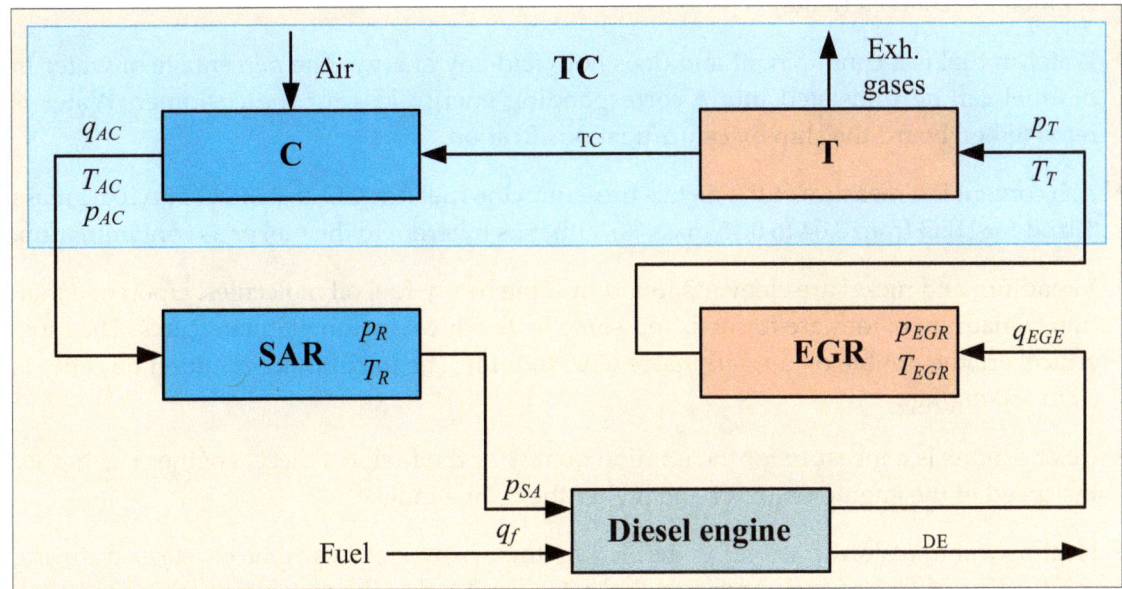

Figure 6. Principal scheme of turbocharging marine diesel engine

3.3. Types of fuel oils for marine diesel engine

Marine diesel engines use three types of liquid fuels standardised by ISO 8217 fuel standard for marine distillate fuels as gas oil (GO) with max. density of 890 kg/m³ at 15 °C, marine diesel oil (MDO) with max. density from 890 to 900 kg/m³ at 15 °C and heavy fuel oil (HFO) with max. density from 920 to 1010 kg/m³ at 15 °C as in [13]. ISO 8217 fuel standard for marine distillate fuels also defines the values of the essential properties of each type of the fuel. The most important properties of marine fuels are as follows:

- Kinematic viscosity, expressed as mm²s⁻¹, is a measure for the fluidity of the fuel at 50 °C. The viscosity of a fuel decreases with increasing temperature. The moment the fuel leaves the injectors, the viscosity must be within the limits prescribed by the engine manufacturer in order to obtain an optimal spray pattern; otherwise, it will lead to poor combustion, deposit formation and energy loss.

- Density expressed as kg/m^3 at 15 °C is an indicator of the ignition quality particularly for the low-viscosity HFOs.

- The sulphur content which depends on the crude oil origin and the refining process. When a fuel burns, sulphur is converted into sulphur oxides which has environmental implications. Maximum permitted sulphur content in the marine fuels is 3,5 %.

- Flash point is the temperature at which the vapours of a fuel ignite. The minimum flash point for marine fuels is 60 °C as per the IMO SOLAS Convention.

- Carbon residue is determined by a laboratory test performed under specified reduced air supply. It gives an indication of the amount of hydrocarbons in the fuel which have difficult combustion characteristics.

- Water in fuel is a contaminant and does not yield any energy. The percentage of water in the fuel can be translated into a corresponding energy loss for the customer. Water is removed on board the ship by centrifugal purification.

- Ash content is a measure of the metals present in the fuel (for GO and MDO max. 0,01 mass % and for HFO from 0,04 to 0,15 mass %), either as inherent to the fuel or as contamination.

- Vanadium and nickel are elements found in some heavy fuel oil molecules. Upon combustion, vanadium oxides are formed, and some have critical melting temperatures. The most critical are the double oxides/sulphates with sodium. The maximum permitted quantity is 50 to 450 mg/kg.

- Cetane index is a measure for the ignition quality of the fuel in a diesel engine. The higher the speed of the engine required, the higher the cetane index.

- Heating value (calorific value) is defined as the amount of heat that is released during combustion of 1 kg of fuel. Assumed that after combustion the water content in the fuel is present as vapour, the condensation heat is not included in the heating value and it is referred to as the net calorific value or lower heating value. Standard net calorific value for GO is 44,000 kJ/kg, for MDO 42,000 kJ/kg and for HFO 40,500 kJ/kg.

4. Combustion process in the marine diesel engines and formation of the air pollutants

4.1. Combustion stoichiometry

The aim of the combustion stoichiometry is to determine the required amount of air and fuel in order to achieve complete combustion. A stoichiometric mixture contains the exact amount of fuel and oxidiser, so that after combustion is completed, all the fuel and oxidiser are consumed to form combustion products. This ideal stoichiometric mixture approximately yields the maximum flame temperature, as all the energy released from combustion is used to heat the products. As in references [14], combustion stoichiometry for a general hydrocarbon fuel (CαHβOy) can be expressed by equation (1) and it can be applied only for single-component hydrocarbons (HC):

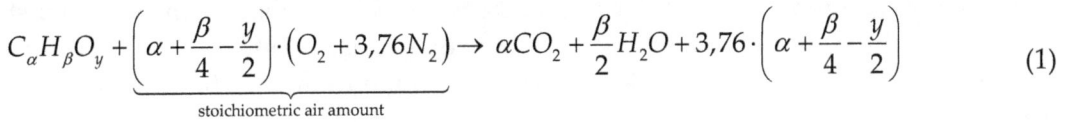

$$C_\alpha H_\beta O_y + \left(\alpha + \frac{\beta}{4} - \frac{y}{2}\right) \cdot \underbrace{\left(O_2 + 3,76N_2\right)}_{\text{stoichiometric air amount}} \to \alpha CO_2 + \frac{\beta}{2} H_2 O + 3,76 \cdot \left(\alpha + \frac{\beta}{4} - \frac{y}{2}\right) \qquad (1)$$

Typical approaches for multiple-component hydrocarbon fuels develop the stoichiometric combustion using the general principle of atomic balance, making sure that the total number of C, H, N and O atoms is the same in the products and the reactants (e.g. multiple-component mixture of a 95 % methane (CH_4) and 5 % hydrogen (H_2)):

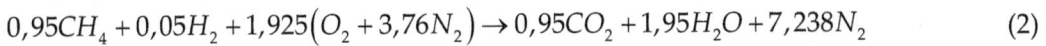

$$0,95CH_4 + 0,05H_2 + 1,925\left(O_2 + 3,76N_2\right) \to 0,95CO_2 + 1,95H_2O + 7,238N_2 \qquad (2)$$

If less air than the stoichiometric amount is used, the mixture is described as rich fuel or rich mixture, and if excess air is used, the mixture is described as lean fuel or lean mixture. For this reason, it is appropriate to determine the amount of the combustible mixture using one of the following methods: a)Fuel-air ratio (FAR), b) equivalence ratio (Φ) and c) percent excess air (% AE).

a. *Fuel-air ratio (FAR)* or f is the actual ratio of fuel mass m_f and air mass m_a and it is expressed as

$$f = \frac{m_f}{m_a} \qquad (3)$$

and it is usually bounded by 0 and ∞. For a stoichiometric mixture, equation (3) becomes

$$f_s = \frac{m_f}{m_a}\bigg|_{ST} = \frac{M_f}{\left(\alpha + \dfrac{\beta}{4} - \dfrac{y}{2}\right) \cdot 4,76 \cdot M_a} \qquad (4)$$

where M_f is the molar mass of fuel and M_a molar mass of air which is approximately 28.96 kg/kmol. The stoichiometric mixture fuel-air ratio of the most hydrocarbon fuels is bounded by 0.05 and 0.07. Air-fuel ratio (AFR) is reciprocal of FAR and it is expressed as AFR=f $^{-1}$.

b. *Equivalence ratio (Φ)* is the actual ratio of fuel-air ratio f to the stoichiometric fuel-air ratio f_s:

$$\Phi = \frac{f}{f_s} = \frac{m_{as}}{m_a}, \qquad (5)$$

and its value is bounded by 0 and ∞. $\Phi < 1$ is a lean mixture; $\Phi = 1$ is a stoichiometric mixture; and $\Phi > 1$ is a rich mixture. The fuel in the combustion process must be mixed with a greater amount of air than in stoichiometric mixture because it is not possible to bring the ideal amount of air to each fuel molecule in order to mix them perfectly so that complete combustion is achieved. In the combustion analysis, an alternative variable lambda (λ) is often used by engineers. Lambda is the ratio of the actual air-fuel ratio to the stoichiometric air-fuel ratio defined as

$$\lambda = \frac{AFR}{AFR_s} = \frac{1}{f/f_s} = \frac{m_a}{m_{as}} = \frac{1}{\Phi} \tag{6}$$

c.   *Percent excess air* (% AE) is the amount of air in excess of the stoichiometric amount and it is defined as

$$\%EA = 100 \cdot \frac{m_a - m_{as}}{m_{as}} = 100 \cdot \left(\frac{m_a}{m_{as}} - 1 \right) \% \tag{7}$$

4.2. Combustion process in marine diesel engine

Process of fuel combustion is comprised of the following steps: entry of fuel jet into the combustion chamber, disintegration of the jet into droplets, decomposition of larger droplets into smaller, droplet heating, droplet evaporation, mixing of fuel vapour with the surrounding air, simultaneous auto-ignition of fuel mixture in several places, continued evaporation of the droplets and burning around (diffusion combustion), formation of soot during combustion in an area near droplets, temperature drop and slowing reaction due to expansion in the cylinder.

While the combustion temperature is still high, it is necessary that the soot particulate finds their reactants (oxygen) to complete combustion reaction. The phases until the simultaneous ignition of fuel mixture represent a delayed auto-ignition in several places which can be defined as the time or engine crank angle that elapses from the beginning of fuel ignition to the auto-ignition of the mixture.

A good spatial distribution of fuel affects the proper and economical operation of the engine. To achieve a good spatial distribution of fuel, it must be injected at a rate of about 150–400 ms^{-1}, which requires a pressure of over 80 MPa.

Dispersion quality is determined by the injection speed, fuel surface tension, fuel viscosity, density of air in the cylinder, turbulence and cavitation in the nozzle. Better turbulence, mixing with air and combustion can be achieved by better penetration and propagation of jet fuel.

In marine diesel engine, injection is performed by injectors with the nozzles that direct the fuel into the cylinder space. Under the influence of aerodynamic forces of compressed air, fuel jet expands and breaks down into small droplets. The quality of fuel atomisation is defined by a mean diameter of the droplets and their uniformity. Better fuel dispersion is achieved with the smaller diameters of the nozzle holes, greater injection pressure and higher compression pressure inside the cylinder. The combustion process in a diesel engine can be divided into four phases (see Figure7).

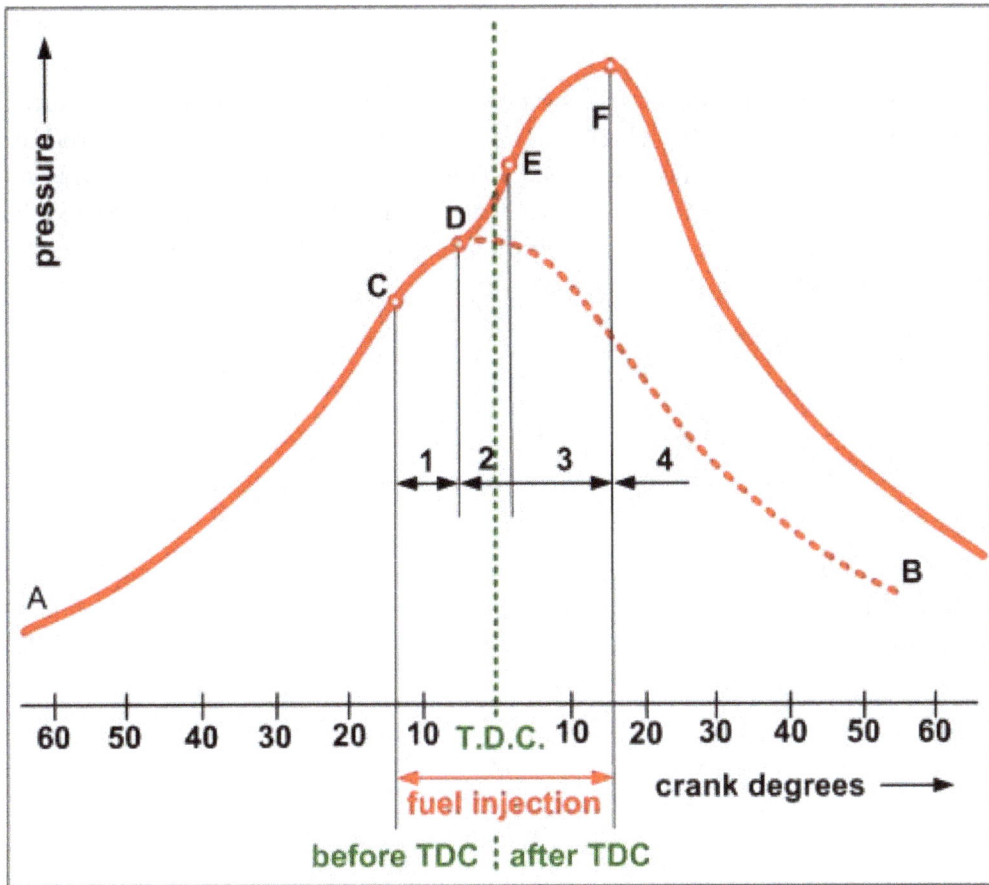

Figure 7. Phases of the combustion process

The first phase, 'ignition delay, curve C-D', defines the period from the beginning of injection until the ignition starts and has an impact on the pollutant formation. This period defines fuel atomisation, evaporation, mixing and the reaction beginning. At sufficiently low turbulence, local flame fronts are created and produce high temperature without soot.

The second phase, 'uncontrolled combustion, curve D-E', is a homogeneous phase of combustion. At this stage, there are sudden ignition and combustion of the already prepared fuel mixture during the delayed ignition phase. Combustion begins simultaneously in several places and conducts intensively, and there is a sudden increase in the pressure and the temperature.

The third phase, 'partially controlled combustion, curve E-F', is diffusion combustion when the fuel droplets vaporise from the surface. Evaporated fuel is mixed with air, and combustion speed is limited by the rate of fuel evaporation and the speed of creating fuel mixture.

The fourth phase, 'after burning, curve F till the end', is the final part of the combustion and it takes about half of the total combustion time duration. During that phase, reaction slows due to expansion and decrease amounts of reactants, and a part of soot that is created during combustion leaves the cylinder as portion of emissions.

4.3. Adiabatic flame temperature

One of the most important features of a combustion process is the highest temperature of the combustion products that can be achieved. The temperature of the combustion products will be the highest when there are no heat losses to the surrounding environment and when all energy released from combustion is used to heat the products. Constant pressure adiabatic temperature calculation, using a mean specific heat capacity method, can be performed for the lean and the rich combustion mixture as in [14]:

a.   *For a lean mixture ($\Phi < 1$):*

$$T_{AFT} = T_R + \frac{\Phi \cdot f_s \cdot LHV}{\left(1 + \Phi \cdot f_s\right) \cdot \overline{c}_{p,P}} \tag{8}$$

where T_R represents the temperature of the reactants, i.e. fuel which has the compression temperature (T_2) after injection and ignition delay, and $\overline{c}_p$ is an average specific heat capacity of the mixture.

b.   *For a rich mixture ($\Phi > 1$):*

$$T_{ATP} = T_R + \frac{f_s \cdot LHV}{\left(1 + \Phi \cdot f_s\right) \cdot \overline{c}_{p,P}} \tag{9}$$

5. Formation of the air pollutants during combustion process in marine diesel engine

The major pollutants in diesel exhaust emissions are a direct result of the diesel combustion process itself. Typical concentrations of exhaust gas emissions from marine diesel engine largely comprise nitrogen approximately 76 %, oxygen abt. 13 %, carbon dioxide (CO_2) abt. 5 % and water vapour abt. 5 %, with smaller quantities of pollutants: nitrogen oxide (NO_x) abt. 1200 ppm, sulphur oxide (SO_x) abt. 640 ppm, carbon monoxide (CO) abt. 60 ppm, partially reacted and non-combusted hydrocarbons (HC) abt. 180 ppm and particulate matter (PM) abt. 120 mg/Nm³ [15]. The composition of this gas mixture, liquids and solids that are actually emitted into the air will vary depending on engine type, engine power, operating conditions as well as fuel and lubricating oil type and also depends on whether the emission control system is present. Pollutant formation in marine diesel engine is discussed below.

5.1. Nitrogen oxides (NOx)

Nitrogen oxides (NOx) generate thermally from atmospheric nitrogen oxygen in the intake or scavenging air. The oxidation of atmospheric nitrogen is influenced by local conditions in the combustion chamber, such as the maximum cylinder pressure, local peak temperatures and local air- fuel ratio. The primary reaction product is nitric oxide (NO) by approximately 90 % of the volume, but about 5 % of it is converted into nitrogen dioxide (NO_2) later in the combustion cycle, during expansion and during the flow through the exhaust system. At the same time, a very limited proportion of nitrous oxide N_2O is also formed. Further oxidation of NO to NO_2 subsequently continues at ambient temperatures after the exhaust gases have passed out to the atmosphere. Nitrogen oxide is of particular concern because of its detrimental effects on respiration and plant life, as well as its significant contribution to acid rain. In addition, NOx, together with volatile organic compounds (VOC), is also involved in a series of photochemical reactions that lead to an increase in troposphere ozone which, in turn, adversely affects human health and natural vegetation. These problems are only pronounced on land and especially in urban areas.

Analysis of the combustion process in the cylinder and the reactions which are involved in formation of NO has identified three main sources of NO formation of which, as mentioned above, some is converted to NO_2 to give the NOx mixture. These sources are thermal NO formation, prompt NO formation and fuel source. A majority of the NO emission is generated by internal combustion engines through the thermal process.

a. *Thermal nitric oxide (NO) formation*

During the combustion process in diesel engine, high temperatures are reached. Around 1700 K, and above up to 2500 K, sufficient thermal energy is available to dissociate oxygen, nitrogen and also other molecules formed during the combustion process itself. The recombination of the elements leads to the formation of NO. The reaction processes are quite slow so that most nitrogen oxides are formed during the mixing of the stoichiometric combustion gases with excess air in the cylinder. In low- and medium-speed diesel engines, by far the most important part of NOx is generated in the thermal NO process.

Formation of nitric oxide can be represented with three chemical reactions based on Zeldovich mechanism as in [16]:

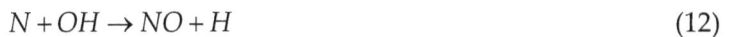

$$N_2 + O \rightarrow NO + N \tag{10}$$

$$N + O_2 \rightarrow NO + O \tag{11}$$

$$N + OH \rightarrow NO + H \tag{12}$$

The first two reactions show the formation of nitric oxide for the lean mixture and the third for the rich mixture. The first reaction is the rate-limiting step due to its very high temperature activation. The high activation energy is required to break the triple bond in the nitrogen molecule (:N≡N:), which occurs at high combustion temperature; this is named thermal nitrogen monoxide (NO). The formation rate of thermal NO is practically insignificant if the temperature is below 1700 K. On the other hand, if the temperature rises, especially over 2000 K, the formation of thermal NO is strongly accelerated. The formation of thermal NO may be reduced by lowering and controlling the temperature peaks and minimising flue gas residence at high temperatures. As in [17], the equation for the total formation rate of thermal nitrogen oxides (NOx) is

$$\frac{d\left[NO_x\right]}{dt} = \frac{6 \cdot 10^{16}}{T^{0,5}} \cdot e^{\left(\frac{-69090}{T}\right)} \cdot \left[N_2\right] \cdot \left[O_2\right]^{0,5}, \tag{13}$$

where T is absolute flame temperature (K), N_2 nitrogen molecule concentration (molcm^{-3}), O_2 oxygen molecule concentration (molcm$^-$) and dNOx/dt nitrogen oxide speed formation (molcm^{-3}).

b.   *Prompt nitric oxide (NO) formation*

Prompt nitric oxide can be formed promptly at the flame front by the presence of hydrocarbon radicals produced only at the flame front at relatively low temperature. Nitric oxide generated via this route is named 'prompt nitric oxide (NO)'. Hydrocarbon (HC) radicals react with nitrogen molecules with the following sequence of reaction steps:

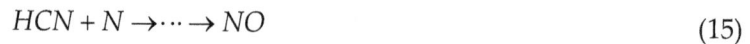

$$CH + N_2 \rightarrow HCN + N \tag{14}$$

$$HCN + N \rightarrow \cdots \rightarrow NO \tag{15}$$

Nitrogen reacts with an HC radical to produce hydrogen cyanide HCN, and further, HCN reacts with nitrogen to produce nitric oxide via a series of intermediate steps. In contrast to thermal NO mechanisms that have activation temperature above 1700 K from (160), prompt NO can be formed starting at low temperature, around 1000 K as in [17].

c.   *Fuel sources of NO formation*

NO formation from fuel becomes important when using heavy fuel oil because such fuels contain more organic nitrogen than marine diesel oil and other distillate fuels. Heavy fuel oil can contain up to 0.5 % nitrogen which increases the total NOx emission by as much as 10 %.

5.2. Sulphur oxides (SOx)

Formation of sulphur oxides (SOx) in the exhaust gases is caused by the oxidation of the elemental sulphur in the fuel into sulphur monoxide (SO), sulphur dioxide (SO_2) and sulphur trioxide (SO_3) during the combustion process. SOx emissions in diesel engine exhaust gas mostly comprise of sulphur dioxide and a small amount of sulphur trioxide. The stable products such as sulphur dioxide (SO_2), hydrogen sulphide (H_2S), carbon disulfide (CS_2) and disulfide (S_2) are created during the combustion of the rich mixtures. The radical sulphur monoxide (SO) reacts with oxygen (O_2) to produce sulphur dioxide (SO_2) at high temperatures. The amount of sulphur dioxide emissions depends on the sulphur content of the fuel used and cannot be controlled by the combustion process. Furthermore, sulphur trioxide (SO_3) cannot be created in the combustion under fuel-rich conditions, even when the combustion is near the stoichiometric point. However, if there is even a 1 % air excess, sulphur trioxide rapidly increases in its quantity. Typically, the amount of SO_3 is 5 % of the amount of sulphur oxides (SO_2 and SO_3). For example, if the fuel contains 3 % sulphur, the volume of SOx generated is around 64 kg per tonne of fuel burned; if fuel with 1 % sulphur content is used, SOx emission amount is about 21 kg per tonne of fuel burned as in [18]. SOx formed from diesel exhaust is corrosive and is partly neutralised by an engine's lubricating oil which is used as a typical base. Moreover, sulphur oxides (SOx) combine with moisture to form sulphuric acid (H_2SO_4), which is then excreted in the form of acid rain. It has a harmful effect on plants and human health and can damage many objects including buildings. Sulphur dioxide emissions also negatively impact human health; sulphate particles particularly can induce asthma, bronchitis and heart disease.

5.3. Carbon monoxide (CO)

The formation of carbon monoxide (CO) is a result of incomplete combustion of organic material, which is due to a lack of oxygen or low temperature at some points in the combustion chamber. Also the same reasons lead to the formation of hydrocarbons (HC). Hydrocarbons can also be formed from evaporation of the lubrication oil towards the end of the firing period. In diesel engines, the formation of carbon monoxide is determined by the air/fuel mixture in the combustion chamber, and since diesel fuel has a consistently high fuel-air ratio and the efficient combustion process, formation of this toxic gas is minimal. Nevertheless, insufficient combustion can occur if the fuel droplets in a diesel engine are too large or the level of turbulence is insufficient or swirl is created in the combustion chamber. When burning heavy fuel oil, the hydrocarbon emissions are lower than from the light fuel oil combustion due to lower evaporating level.

5.4. Hydrocarbon (HC)

Hydrocarbon (HC) emissions as fraction of the exhaust gases from diesel engines predominantly consist of unburned or partially burned fuel and lubricating oils as a result of insufficient temperature. This often occurs near the cylinder wall where the temperature of the air/fuel mixture is significantly less than in the centre of the cylinder. In the atmosphere, the hydrocarbons are subjected to photochemical reactions with nitrogen oxides forming the ground-level ozone and smog. Hydrocarbon (HC) emissions are represented as total hydrocarbons (THC) or as non-methane hydrocarbons (NMHC), as in [19].

5.5. Particulate matter (PM)

Particulate matter is a mixture of organic and inorganic substances largely comprising elemental carbon, ash minerals, heavy metals, condensed sulphur oxides, water and a variety of unburned or partially burned hydrocarbon components of the fuel and lubricating oils. More than half of the total particulate mass is soot (inorganic carbonaceous particles), of which the visible evidence is smoke. Some of the fuel particles do not burn completely, and they are emitted as droplets of heavy liquid or carbonaceous material. The incomplete burning is a result of locally low quantities of excess air. A mistimed or otherwise poorly operating fuel injection and poor mixing of fuel within the cylinder also result in incomplete combustion and increased the particulate matter emissions. Soot particles (unburned – elemental carbon) are not themselves toxic, but they can cause the build-up of aqueous hydrocarbons (HC), and some of them are believed to be carcinogens. Particulates constitute no more than around 0.003 % of the engine exhaust gases. Almost the entire diesel particle mass is in the fine particle range of 10 microns or less in diameter (PM_{10}). Approximately 94 % of the mass of these particles are less than 2.5 microns ($PM_{2.5}$) in diameter. Diesel PM is of specific concern because it poses as a lung cancer hazard for humans as well as a hazard from noncancer respiratory effects such as pulmonary inflammation. Because of their small size, the particles are readily respirable and can effectively reach the lowest airways of the lung along with the adsorbed compounds, many of which are known or suspected mutagens and carcinogens. Secondary reactions of NOx and SOx can also produce PM.

The most effective method of reducing particulate emissions is to use lighter distillate fuels; however, this leads to added expense. Additional reductions in particulate emissions can be achieved by increasing the fuel injection pressure to ensure that optimum air/fuel mixing is achieved. The third method of reducing particulate emissions is to use cyclone separators, which are effective for particle sizes greater than 0.5 μm.

5.6. Carbon dioxide (CO_2)

Carbon dioxide is one of the basic products of combustion and is not toxic; however, it has been linked to the 'greenhouse effect' and global warming. Diesel engine exhaust gases containing CO_2 as a result of carbon and oxygen O_2 combustion. The maximum concentration of carbon dioxide will be generated during stoichiometric combustion, i.e. when complete amount of fuel reacts with oxygen from the air during combustion. The actual concentration of CO_2 depends on the relative contents of carbon (C), hydrogen (H) and other combustible elements in the fuel. The maximum values of carbon dioxide for common types of marine fuel are shown in Table 3 [20], assuming that the exhaust is dry.

The maximum value of carbon dioxide (CO_2 max) can be calculated according to the following expressions:

$$CO_2\ max = \frac{\text{No. of } CO_2 \text{ molecules produced by complete combustion of fuel}}{\text{Total no. of molecules of combustion products}} \quad (16)$$

Fuel	CO_2 max (%)
Natural gas	11.9
Light fuel oil	15.5
Heavy fuel oil	15.8

Table 3. CO_2 max values for marine fuel, assuming the gases are dry

For 'wet' exhaust gases

$$CO_2 \max = \cfrac{c}{c + \cfrac{h}{2} + \cfrac{79,1}{20,9} \cdot \left(c + \cfrac{h}{4} \right)} \ \%, \tag{17}$$

For 'dry' exhaust gases

$$CO_2 \max = \cfrac{c}{c + \cfrac{79,1}{20,9} \cdot \left(c + \cfrac{h}{4} \right)} \ \% \tag{18}$$

Carbon dioxide (CO2) concentration can be calculated in the exhaust gas emissions according to equation (19), provided that oxygen concentration (O_2), maximum concentration of carbon dioxide (CO_2) max and fuel type are known, as in [20]:

$$\left[CO_2 \right] = \frac{CO_2 \max \cdot \left(20,9 - \left[O_2 \right] \right)}{20,9} . \tag{19}$$

Emission	Source
NOx	Function of peak combustion temperature, oxygen content and residence time (function of engine speed in rpm)
SOx	Function of fuel oil sulphur content
CO	Function of the air excess ratio, combustion temperature and air/fuel mixture
HC	Function of the amount of fuel and lubricating oil left unburned during combustion process
PM	Originates from unburned fuel as well as ash content in fuel and lubricating oil
CO_2	Function of combustion

Table 4. Summary of pollutants

Reduction of carbon dioxide emissions can be achieved by reducing specific fuel oil consumption (SFOC) since the amount of CO_2 produced is directly proportional to the volume of fuel used and therefore to the engine efficiency. An alternative is to use fuel with a low carbon ratio relative to hydrogen, which greatly increases the price of marine fuel oils. Table 4 provides a summary for the pollutants discussed above.

6. Methods for reducing harmful pollutant emissions from marine diesel engines

Different methods of reducing the pollutant emission from ship's diesel engines are briefly described in the following text. The focus is more on NOx and SOx than on other emissions, but some attention will be given to them as well.

6.1. NOx emission reduction technologies

Nitrogen oxides (NOx) are an important air pollutant created as a by-product of combustion. Air contains primarily nitrogen (N_2) and oxygen (O_2). The heat generated during combustion causes nitrogen (N_2) and oxygen (O_2) to react to form NO_X which is in direct proportion to peak combustion temperature and pressure. Therefore, NO_X emissions can be mitigated by engine controls that decrease combustion temperature and/or aftertreatment of the exhaust gas. NOx reduction methods are generally categorised as primary methods or internal measure and secondary methods or aftertreatment. The primary methods include changes to the combustion process within the engine and can be divided into three main categories: combustion optimisation, water-based controls and exhaust gas recirculation. Many of the mentioned methods aim to reduce NOx emissions by reducing peak temperatures and pressures of the combustion process in the engine cylinder. Secondary methods or aftertreatment implies post-combustion abatement in which the exhaust gas is treated in order to remove NO_X, either passing it through a catalyst or plasma system. Each of these methods is discussed below.

6.1.1. Primary methods or internal measure

The primary methods include changes to the combustion process within the engine and can be divided into three main categories. The first category is combustion optimisation; the second one is water-based control which consists of water injection, water/fuel emulsion and humidification; and the third category is exhaust gas recirculation. These methods have generally low impact on fuel consumption. Another drawback is to retrofit the system of existing engines. Tier II limits under MARPOL Annex VI, Regulation 13, can be achieved using primary controls.

a.   *Combustion optimisation*

There are a number of ways to modify the combustion process, each aimed to reduce NOx emissions. Optimisation of the engine combustion process includes modifying the spray pattern by modification of the fuel valve design, injection timing, intensity of injection and injection rate profile (injection rate shaping), compression ratio, scavenge air pressure and scavenge air cooling. Delayed injection timing is very effective in reducing NOx but increases fuel consumption and smoke. It is usually combined with increased compression pressure and decreased injection duration to minimise or avoid increase in fuel consumption. Other operational modifications that could be made to reduce emissions are combustion chamber optimisation, variable valve timing, increasing the turbo efficiency, the use of a fuel injection system that can be easily adjusted (e.g. electronically controlled injection system) and decrease in the engine air intake temperature using Miller supercharging.[4] Modification of the fuel valve design means replacing the conventional injectors with fuel efficient valves (e.g. slide valves) that optimise the fuel injected into the cylinder. These valves differ from conventional valves in their spray patterns, and they are designed to reduce the dripping of fuel from the injector into the postinjection combustion zone. This fuel entering late the combustion zone is subjected to lower temperatures and therefore results in the emission of unburnt fuel (PM) and VOCs. Changing the conventional fuel valves with slide valves has a significant impact on NOx reduction and PM emissions since PM is a product of incomplete combustion and unburnt fuel. Currently, the slide fuel valves are only applicable for slow-speed two-stroke diesel engines. However, all new engines of this type are supposed to have these valves fitted as the standard. The fuel nozzle was optimised for NOx simultaneously with the development of the slide valve. Tests on a 12K90MC engine (55MW at 94 rpm) at 90 % load showed a 23 % reduction in NOx emissions for a slide-type valve compared with a standard valve and nozzle and with a 1 % fuel consumption increase. Furthermore, increasing the number of injectors per cylinder enables the combustion process to be better controlled and therefore more efficient combustion. However, additional injectors, piping and associated equipment are associated with a cost penalty. Nowadays, modern slow-speed engines use three fuel injectors located near the outer edge of the combustion chamber. With sequential injection, each of the three nozzles in a cylinder is actuated with different timings. Pulsed injection gives about 20 % NOx reduction with about 7 % increase in fuel consumption. Sequential and pre-injection gave less NOx reduction and less fuel consumption increase. The effects are the result of changes in the overall pressure development and interaction between fuel sprays. Pre-injection can be used to shorten the delay period in medium-speed engines and thus decreases temperature and pressure during the early stages of combustion, resulting in reduced NOx. Pre-injection can reduce particulates which are increased by other NOx control measures, thus allowing greater flexibility in NOx control. Delayed injection combined with increased compression ratio have effect on reducing the maximum combustion pressure and hence temperature. By using this simple technique, a reduction of up to 30 % can be achieved. However, delayed injection increases fuel consumption up to 5 % in specific fuel oil consumption due to later burning, as less of the combustion energy release is subjected to the full expansion process and gas temperatures remain high later into the expansion stroke, resulting in more heat losses on the

4 Reduced scavenge air temperature reduces combustion temperatures and thus NOx. For every 3 OC reduction, NOx may decrease by about 1 %. On four-stroke engines, the Miller concept can be applied to achieve low scavenge air temperature. Using a higher-than-normal pressure turbocharger, the inlet valve is closed before the piston reaches bottom dead centre on the intake stroke. The charge air then expands inside the engine cylinder as the piston moves towards bottom dead centre, resulting in a reduced temperature. Miller supercharging can reduce NOx by 20 % without increasing fuel consumption.

walls. In some engines, the timing adjustment can be made while in service. Smoke and emission of PM also increase due to reduced combustion temperatures and thus less oxidation of the soot produced earlier in the combustion, as in [21]. Furthermore, increasing injection pressure leads to better atomisation of the fuel and therefore to reduction in particulates and CO. Since combustion is cleaner, this technique tends to increase NOx reduction. There is also the new generation of the electronically controlled camshaftless engines that allow great flexibility for optimisation of the combustion process over the full range of operating conditions. These computer-controlled engines have allowed greater operational flexibility. As far as NOx is concerned, the main features are computer control of variable injection timing (VIT), injection rate shaping, variable injection pressure and variable exhaust valve closing (VEC). Variable exhaust closing gives the ability to change the effective compression ratio. With variable injection timing and variable exhaust valve closing, it is possible to optimise the injection timing delay and increased compression ratio over the whole load range to maintain peak pressures at low load while avoiding excessive peak pressures at high load. Computer-controlled camshaftless engines are equipped with common rail injection techniques which give high injection pressures and thus good spray characteristics even at low loads granting of NOx reduction emissions.

b. Water-based control

The second category is water-based controls consisting of water injection, water/fuel emulsion and humidification. Water-based controls reduce emissions of the NOx from diesel engines by introducing freshwater at different stages of the combustion process. Introduction of water into the combustion chamber reduces maximum combustion temperature due to the absorption of energy for evaporation and the increase in the specific heat capacity of the cylinder gases and thus reduce emissions of the NOx [23]. The in-cylinder evaporation of the water also improves the atomisation of the fuel and causes it to burn more completely. Freshwater can be introduced in the charge air (humidification), through direct injection into the cylinder or through water/fuel emulsion. Water/fuel emulsion is the process of introducing water into the fuel prior to injection into the combustion cylinder and can reduce smoke, while humidification can increase smoke. Direct water injection is the process of introducing water directly into the combustion cylinder at pressures of 200–400 bar. The water is injected into the cylinder by a combined injection valve and nozzle that allow injection of water and fuel oil. The process is electronically controlled. Direct injection of water and water/fuel emulsions place the water more directly in the combustion region, where it has maximum effect on NOx production. Generally, direct water injection or water/fuel emulsions will yield about 1 % reduction in NOx for every 1 % of water-to-fuel ratio. This one-to-one ratio is consistent up to about 30 % water content, at which point the combustion temperature decreases too much, resulting in an increase in PM emissions. Alternative to water injection and water/fuel emulsion is the scavenge air humidification as a second category of the primary methods for NOx emission reduction that implies injection of very fine water mist in scavenge airstream after the turbocharger using special nozzles (Scavenge Air Saturation System). The fine water droplets evaporate fast, and further heat is introduced in the scavenge air. Humidification requires about twice as much water for the same NOx reduction compared with direct injection of water and water/fuel emulsions. Humidification can reduce NOx levels down to 2 to 3 g/kWh without fuel consumption penalty. Similar technique that is used for humidification of the engine scavenge air is the so-called humid air motor (HAM). In this system, hot compressed air from the turbocharger is led to a humidification tower and exposed to a large surface area and flushed with hot water. The water can be heated by a heat exchanger connected to the jacket cooling system or using an exhaust gas boiler. One manufacturer claims considerable success in service in reducing NOx emissions with the added claim of increasing the indicated power of the engine at certain loads, therefore reducing fuel consumption hence proportionally reducing CO_2 emissions. The actual degree of NOx reduction varies from 10 % to over 60 %, depending on the engine type and which of the above reduction methods are adopted. For example, the experiment carried out on the Viking Line's MS Mariella has shown a NOx emission reduction from 17 to between 2.2 and 2.6 g/kWh and a decrease in fuel consumption of 2–3% using the HAM system [24].

c.   *Exhaust gas recirculation (EGR)*

The exhaust gas recirculation (EGR) system is based on lowering of the combustion temperature and oxygen concentration thus lowering NOx. EGR reduces combustion temperatures by increasing the specific heat capacity of the cylinder gases and by reducing the overall oxygen concentration, taking away a part of the exhaust gases and mixing it into the engine intake air. Some of the exhaust gas is cooled and cleaned before recirculation to the scavenge air side. The usage of the exhaust gas as intake air reduces the oxygen content in intake air from 21 % to 13 % which limits the NOx that can be formed and reduces the amount of combustion products that can take place. In engines operating on poor-quality fuel, external EGR can lead to fouling and corrosion problems. The residue from cooling and cleaning the exhaust gas on ships using heavy fuel oil contains sulphur in a form which is difficult to dispose of. The relative changes in measured emission parameters as a function of the recirculation amount at 75 % engine load show that at increased recirculation amounts, the HC and PM emissions are reduced corresponding to the reduction of the exhaust gas flow from the engine. Increased recirculation amount leads to increase in CO emissions due to lower cylinder excess air ratios and thus lack of oxygen in the combustion chamber. Furthermore, EGR tends to increase smoke by reducing the O_2 concentration, increasing the combustion duration and decreasing the combustion temperature. All of that may be controlled using additional techniques such as water in fuel to achieve an optimum balance between NO_x, CO and PM. Test engine work by MAN Diesel & Turbo has shown that, with 40 % recirculation, EGR has the potential to reduce NO_x down to Tier III levels on a two-stroke low-speed marine engine and that increased fuel consumption, carbon monoxide emissions and PM emissions resulting from reduced combustion efficiency are manageable with engine adjustments. It is also reported that specific fuel consumption is greatly improved when using EGR to reduce NO_x down to Tier II limits, when compared with using engine adjustments to achieve the same level of emissions, particularly at part load as in [25]. There are many different components to an EGR system such as high pressure exhaust gas scrubber fitted before the engine turbocharger, cooler to further reduce the temperature of the recirculated gas, water mist catcher to remove entrained water droplets, high-pressure blower to increase recirculated gas pressure before reintroduction to the engine scavenge air and automated valves for isolation of the system. The scrubber in the EGR system is used to remove sulphur oxides and particulate matter from the recirculated exhaust, to prevent corrosion and reduce fouling of the EGR system and engine components. The system requires the use of an electrostatic precipitator and catalysts to remove the particulates from the exhaust gas before injecting it as intake air. There is also a need for wet-scrubbing technology to remove the sulphur components of the exhaust stream prior to reintroduction into the engine. A cooling unit is also needed to reduce the temperature of the exhaust gas before it returns to the engine.

6.1.2. Secondary methods or aftertreatment

Secondary methods, or aftertreatment, are based on treating the engine exhaust gas itself by passing it through a catalyst or plasma system. There has been much development in selective catalytic reduction (SCR) and nonthermal plasma (NTP) reduction systems over the last few years. Using these methods, NOx emission reductions of over 95 % can be achieved [26].

Selective catalytic reduction (SCR) is an exhaust gas treatment method by which the NOx generated in a marine diesel engine exhaust gas can be reduced to a level in compliance with the NOx Tier III requirements. The method involves mixing of ammonia as a reducing agent with the exhaust gas which is passed over a catalyst where more than 90 % of the NOx can be removed to below 2 g/kWh. The SCR system converts nitrogen oxides into harmless nitrogen and water, by means of a reducing agent injected into the engine exhaust stream before a catalyst. Hydrocarbons are also reduced. Exhaust emission abatement systems using SCR technologies usually use an ammonia (NH_3) reductant introduced as a urea/water solution (($(NH_2)_2CO$)) into the exhaust stream, prior to the catalyst blocks. For marine systems, a 40 % solution of urea in de-ionised water is typically used for safe handling and toxic risk reasons [27, 28]. The use of urea in the system breaks down the NOx emissions to N_2 and H_2O. The degree of NOx removal depends on the amount of ammonia added. A NOx reduction efficiency of 90 % can be achieved using a urea injection rate of 15 g/ kWh. NOx is reduced according to the following overall reaction scheme [26,27,29]:

Urea decomposition before entering the reactor:

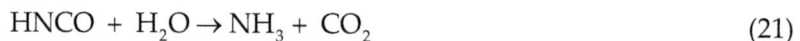

$$(NH_2)_2CO \left(urea\right) \rightarrow NH_3 \left(ammonia\right) + HNCO \left(isocyanic\ acid\right) \tag{20}$$

$$HNCO + H_2O \rightarrow NH_3 + CO_2 \tag{21}$$

The resulting quantity of CO_2 is minor when compared with that resulting from fuel combustion.

NO_x reduction at the catalyst:

$$4NO + 4NH_3 + O_2 \rightarrow 4N_2 + 6H_2O \tag{22}$$

$$2NO + 2NO_2 + 4NH_3 \rightarrow 4N_2 + 6H_2O \tag{23}$$

$$6NO_2 + 8NH_3 + O_2 \rightarrow 3N_2 + 6H_2O \tag{24}$$

Equation 22 shows the main SCR reaction as nitric oxide dominates in the exhaust. The reaction shown at equation 23 occurs at the fastest rate up to an $NO_2:NO$ ratio of 1:1. However, at higher ratios, the excess NO_2 reacts slowly as per equation 24. The rate of urea injection must be sufficient to reduce NO_X emissions to the required level but not so great to avoid ammonia slip. Control is based on the load and speed of the engine with active feedback provided on some systems by NO_X and ammonia emission monitoring. At engine start-up, urea injection is initiated once the catalyst reaches operating temperature, which is key for effective NO_X reduction performance, deposit prevention and avoidance of ammonia slip. Catalysts have considerable heat capacity so the time taken to reach the injection trigger temperature is dependent on a number of factors including the minimum catalyst operating temperature recommended for the fuel type, the period of cooldown since the engine was last operated, the size of the catalyst and the engine load pattern at start-up. Injection can begin up to 30 min after a fully cold start, whereas it may begin within 10–15 min if the catalyst is still warm from running in the previous 6–10 h. SCR units are typically installed in the exhaust system of a diesel engine, if applicable, before the exhaust gas economiser and as close as possible to the engine because of the relatively high exhaust gas temperatures required by the catalysts for effective NOx reduction reactions. The SCR catalysts may also be integrated with the engine by close coupling to the engine, typically applicable to small high-speed diesel engines. For slow-speed diesel engines with inherently low relative exhaust gas temperatures, this may necessitate the integration of the SCR reaction chamber and catalysts before the turbocharger exhaust turbine. Depending on the engine load, the exhaust gas temperature on this side is 50–175 °C higher than on the low pressure side. Even though the reactor is placed before the turbine, the exhaust gas temperature will normally still be too low at low loads. To increase the temperature, a cylinder bypass from the scavenge air receiver to the turbine inlet is installed. The bypass is controlled by the cylinder bypass valve.

When opening the bypass, the mass of air through the cylinders will be reduced without losing the scavenge air pressure, and, accordingly, the exhaust gas temperature will increase. This system makes it possible to keep the temperatures above the required level. However, the cylinder bypass will increase the SFOC depending on the required temperature increase. Selective catalytic reduction is the only technology currently available to achieve compliance with the Tier III NOx standards for all applicable engines. Another option is selective non-catalytic reduction (SNCR), which works in a similar way with selective catalytic reduction but without the use of a catalyst. A reducing agent (ammonia or urea) injected during the combustion process converts the nitrogen oxides into nitrogen and water, reducing NOx emissions by 50 % [30]. The drawback of this system is that it is less efficient than the SCR method, because only 10–12 % of ammonia reacts with NOx. Since the cost of ammonia is relatively high and since the system requires extensive modification to the engine, the SNCR option does not appear to be competitive. Plasma reduction systems are based on the use of plasma. This is a partially ionised gas composed of a charge of a neutral mixture of atoms, molecules, free radicals, ions and electrons. Electrical power is converted into electron energy, and the electrons create free radicals, which destroy pollutants in exhaust emissions. Experiments have shown that plasma reduction systems can reduce NOx by up to 97 % [30]. It seems to be flexible in terms of size and shape and should be at relatively low cost. However, for marine use, it is still in the development phase.

6.2. SOx reduction technologies

The emission of sulphur dioxide is directly proportional to the content of sulphur in fuel. To meet the restrictions on emissions of sulphur oxides (SOx) and PM (particulate matter) that are determined by the MARPOL convention Annex VI which specifies a global and a local (ECA) limit on the sulphur content in marine fuel, there are only two possibilities for reducing SOx emissions: either use fuels with low sulphur content (there will be no restrictions in the use of heavy fuel oil) or apply an exhaust gas cleaning system to reduce the total emission of SOx. Although the SOx requirements can be met by using a low-sulphur fuel, the regulation allows alternative methods to reduce the emissions of SOx to an equivalent level. The process of exhaust gas cleaning is performed in a scrubber unit. There are two main types of SO_x scrubber: wet scrubbers that use water (seawater or fresh) as the scrubbing medium and dry scrubbers that use a dry chemical. The wet scrubbing technology is based on the fact that sulphur oxides dissolve in water. This means that when the exhaust gas is sprayed with the alkaline water in the scrubber, the SOx will dissolve in the scrubbing water and be cleaned from the exhaust gas. The water is injected into the exhaust gas stream and is discharged from the bottom of the scrubber. The alkalinity in the scrubbing water will neutralise the SOx emissions. The scrubbing water must be cleaned of particulate matter and other contaminants before being discharged out into sea. Wet scrubbing systems which are normally fitted on marine engines may be categorised as open-loop system, closed-loop system or hybrid system.

Open-loop systems – In an open-loop technology, the water comes from the sea and goes directly to the scrubbers. After the scrubbing process, the water goes through water treatment and to the sea again. This system takes advantage of the natural alkalinity of seawater to buffer the acidity of SO_x gases. The seawater flow rate in open-loop systems is approximately 45 m^3/ MWh. Sulphur oxide removal rate is close to 98 % with full alkalinity of the seawater, meaning emissions from a 3.5 % sulphur fuel will be the equivalent of those from a 0.10 % sulphur fuel after scrubbing [29]. The sulphur oxides generated in the combustion process are dissolved and removed by the scrubber water. Sulphur dioxide (SO_2) is dissolved and ionised to bisulphite and sulphite, which is then readily oxidised to sulphate in seawater containing oxygen. As in reference [29], similarly sulphuric acid, formed from SO_3, and hydrogen sulphate dissociate completely to sulphate according to chemical reactions:

For SO_2:

$$SO_2 + H_2O \rightarrow H_2SO_3 \left(sulphurous\ acid\right) \rightarrow H^+ + HSO_3 \left(bisulphite\right) \tag{25}$$

$$HSO_3^- \left(bisulphite\right) \rightarrow H^+ + SO_3^{2-} \left(sulphite\right) \tag{26}$$

$$SO_3^{2-} \left(sulphite\right) + 1/2\,O_2 \rightarrow SO_4^{2-} \left(sulphate\right) \tag{27}$$

For SO_3:

$$SO_3 + H_2O \rightarrow H_2SO_4 \left(sulphuric\ acid\right) \tag{28}$$

$$H_2SO_4 + H_2O \rightarrow HSO_4^- \left(hidrogen\ sulphate\right) + H_3O^+ \tag{29}$$

$$HSO_4^- \left(hidrogen\ sulphate\right) + H_2O \rightarrow SO_4^{2-} \left(sulphate\right) + H_3O^+ \tag{30}$$

Closed-loop systems use freshwater treated with sodium hydroxide (NaOH) as the scrubbing media for the neutralisation of SO_X. This results in the removal of SO_X from the exhaust gas stream as sodium sulphate according to the following chemical reactions as in [29]:

For SO_2:

$$NA^+ + OH^- + SO_2 \rightarrow NaHSO_3 \left(aq\ sodium\ bisulphite\right) \tag{31}$$

$$2NA^+ + 2OH^- + SO_2 \rightarrow Na_2SO_3 \left(aq\ sodium\ sulphite\right) + H_2O \tag{32}$$

$$2NA^+ + 2OH^- + SO_2 + 1/2\,O_2 \rightarrow Na_2SO_4 \left(aq\ sodium\ sulphate\right) + H_2O \tag{33}$$

For SO_3:

$$SO_3 + H_2O \rightarrow H_2SO_4 \left(sulphuric\ acid\right) \tag{34}$$

$$2NaOH + H_2SO_4 \rightarrow Na_2SO_4 \left(aq\ sodium\ sulphate\right) + 2H_2O \tag{35}$$

In a closed-loop technology, absolutely no water comes from the sea. The freshwater comes from a buffer tank and is cooled by the seawater. The freshwater is composed of NaOH and leaves the buffer tank to go to the scrubber. After the scrubbing process, the water comes back to the buffer tank, cleaned by a filter. The black water goes to a sludge tank and the clean water goes back to the scrubbing cycle. A big storage tank fills up the buffer tank. Closed-loop systems can also be operated when the ship is operating in enclosed waters where the alkalinity would be too low for open-loop operation. Closed-loop systems typically consume sodium hydroxide in a 50 % aqueous solution. The dosage rate is approximately 15 l/MWh of scrubbed engine power of a 2.70 % sulphur fuel is scrubbed to equivalent to 0.10 %. Using a closed-loop technology can have some advantages. First, there is a possibility to increase the pH level in order to reduce more SOx. Also there is no corrosion of the parts and less discharge water to clean. The running costs of the closed-loop technology are relatively high because it uses NaOH which is 0.2 €/kg and its required monitoring units. Also the sludge tanks have to be discharged at the harbour which costs a lot of money. The closed-loop reduces about 98 % SOx in the exhaust gas.

Hybrid system – A hybrid system is a mixture of both open loop and closed loop. In harbours and ECA, the system can operate with freshwater without generating any significant amount of sludge to be handed at port calls. At open sea, the system switches to the seawater open loop. Using a hybrid technology can have some advantages. First, if the ship is running at open sea, after switching to open loop, the accumulated water of the buffer tank can slowly be removed back to the sea. Also, the tank is slowly filled up again to prepare for the arrival at sensitive areas.

Dry SO$_X$scrubber known as an 'absorber' brings the exhaust gas from diesel engine in the multistage absorber where contact with calcium hydroxide (Ca(OH)$_2$) granules reacts with sulphur dioxide (SO$_2$), forming calcium sulphite as in reference [29]:

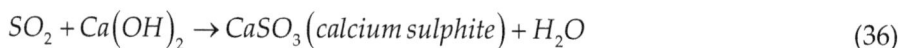

$$SO_2 + Ca(OH)_2 \rightarrow CaSO_3 \left(calcium\, sulphite\right) + H_2O \tag{36}$$

The sulphite is then oxidised and hydrated in the exhaust stream to form calcium sulphate dihydrate, or gypsum:

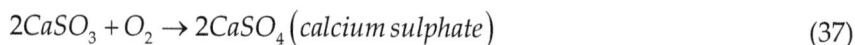

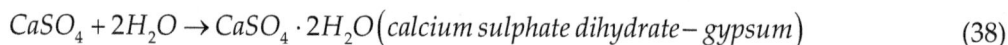

$$2CaSO_3 + O_2 \rightarrow 2CaSO_4 \left(calcium\, sulphate\right) \tag{37}$$

$$CaSO_4 + 2H_2O \rightarrow CaSO_4 \cdot 2H_2O \left(calcium\, sulphate\, dihydrate- gypsum\right) \tag{38}$$

Similarly, chemical reactions take place for SO$_3$:

$$SO_3 + Ca(OH)_2 + H_2O \rightarrow CaSO_4 \cdot 2H_2O \left(calcium\, sulphate\, dihydrate- gypsum\right) \tag{39}$$

Using a dry scrubber can have some advantages. First, the good point of this technology is that the desulphurisation unit requires, aside from electrical energy, only $Ca(OH)_2$ in the shape of spherical granulates. Also the dry scrubber further operates as a silencer. Dry scrubbers typically operate at exhaust temperatures between 240 °C and 450 °C. Calcium hydroxide granules are between 2 and 8 mm in diameter with a very high surface area to maximise contact with the exhaust gas. Within the absorber, the calcium hydroxide granules ($Ca(OH)_2$) react with sulphur oxides to form gypsum ($CaSO_4\ 2H_2O$). To reduce SO_X emissions to those equivalent to fuel with a 0.1 % sulphur content, a typical marine engine using residual fuel with a 2.70 % sulphur content would consume calcium hydroxide granules at a rate of 40 kg/MWh (i.e. a 20 MW engine would require approximately 19 tonnes of granulate per day) [29, 30]. The dry scrubber reduces up to 99 % SOx in the exhaust gas. It will be absolutely no problem to fulfil all the IMO 3 requirements for 2016. The dry scrubber compared to the wet scrubber has lower investment costs and higher running costs and requires a lot of space which reduces the benefits. The efficiency of the SOx scrubber systems depends on the sulphur content in the fuel and generally ranges up to 97 %. Anyhow, the efficiency system must be sufficient to achieve a SOx emission level that is equal to or lower than the required limit.

On the other hand, it should be noted that the reduction of both NOx and SOx emissions from marine diesel engines be achieved by replacing conventional fuels with alternative fuels, e.g. liquefied natural gas (LNG). In case of using LNG, NOx emission can be reduced by 60 % and SOx emission by 90–100 %.

There is also another option to use onshore power supply at ports which is especially beneficial for local air quality. In this case, NOx and SOx emissions can be reduced by 90 %, while CO_2 reduction depends on the source of electricity. The total CO_2 emission reduction depends on how the used electricity is produced. In the European Union, the use of shore-side electricity rather than electricity generated by a ship using low-sulphur fuel will cut CO_2 emissions by an average of 50 %.

One of the main benefits of shore connection systems stems from the fact that electricity generated on land by power plants has a smaller adverse impact on the ecosystem than that produced by ship engines. Namely, the main cause of air pollution from ships in ports is the use of auxiliary diesel engines to generate electricity on ships. Furthermore, experiments with wind and solar power, biofuels and fuel cells are ongoing and could be useful in the future to reduce air pollution from ships.

In Table 5, overview of different technologies and their potential for reduction of emissions from marine diesel engines are summarised as in [31].

Category	Technology aimed to reduce	NOx	SOx	PM	CO_2
Engine modification	NOx				
	Optimise combustion	20–40%	0%	25–50%	0%
Water-based control	Direct water injection	50%	0%	0%	0%

Category	Technology aimed to reduce	NOx	SOx	PM	CO$_2$
	Water in fuel (30 % emulsion)	30%	0%	0%	0%
	Scavenge air humidification	30–60%	0%	0%	0%
	Humid air motor	10–70%	0%	40–60%	0%
Exhaust gas recirculation	ERG	20–85%	0%	0%	0%
Aftertreatment	Selective catalytic reduction	90–95%	0%	0%	0%
	SOx	0%	0%	0%	0%
	Scrubber	0%	90–97%	80–85%	0%
(Alternative) fuels	Low-sulphur fuel 2,7 %S to 0,1 %S	0%	97%	20%	0%
	Both NOx and SOx	0%	0%	0%	0%
	LNG	60%	90–100%	72%	0–25%
	Onshore power supply (in harbour only)	97%	96%	89%	Depending on energy source, average 50%

Table 5. Different technologies and their reduction potential

7. Conclusion

Owing to rapidly developed shipping industry and maritime traffic in recent decades, air pollution emissions from ocean-going ships are continuously growing. Exhaust gases from marine diesel engines are the primary source of emission harmful pollutants such as nitrogen oxides (NOx), sulphur dioxide (SO$_2$), carbon monoxide (CO) and particulate matter (PM) which contribute significantly to environmental pollution, especially in port areas that are often located in or near urban areas, and a significant number of people are exposed to these emissions. The increased air pollutant concentrations and deposition have several negative effects. Nitrogen oxide and particulate matter can contribute to many serious health problems and increased morbidity and mortality (especially from cardiovascular and cardiopulmonary diseases). Nitrogen oxides also contribute to the formation of ground-level ozone, which has a harmful effect on plants and human health. Furthermore, sulphur dioxide and nitrogen oxide emissions increase acidification of sensitive forest ecosystems along the coastal areas and have a harmful effect on plants, aquatic animals and infrastructure by accelerating the deterioration process of various materials. Finally, ships are also a source of greenhouse gas, a pollutant which contributes to global warming. Recent studies indicate that the emission of CO_2 by ship corresponds to about 3 % of the global anthropogenic emissions. If things remain the same, by 2020, shipping will have been the biggest single emitter of air pollution especially in areas of the dense maritime traffic such as Europe, North America and East Asia which surprisingly surpasses the emissions from all land-based sources together. Since harmful pollutant emissions from ships have great impact on the human health and the environment, it is required to tighten uniform regulations at the global level, bearing in mind that shipping is inherently international. The International Maritime Organization (IMO) responsible for the safety of life at sea and the protection of the marine environment reacts on NOx, SOx, PM and CO_2 emissions from a ship by adoption of Annex VI of the MARPOL 73/78 Convention, titled 'Regulations for the Prevention of Air Pollution from Ships'. MARPOL Annex VI sets limits on NOx and SOx emissions from ship exhausts and prohibits deliberate emissions of ozone-depleting substances. Furthermore, the IMO marks out Emission Control Areas (ECAs) in cooperation with national governments with more stringent controls on sulphur emissions. These areas currently comprise the Baltic Sea, the North Sea, the English Channel, the US Caribbean Sea and the area outside North America (200 nautical miles). Ships are currently being permitted to burn fuel oils with sulphur content of less than 3.5 % while operating outside an ECA but must ensure that they burn fuel with a sulphur content of less than 1 % while within the sulphur Emission Control Areas. In accordance with EU's marine fuel sulphur directive, the sulphur content in marine gas oil within the territorial waters of an EU member state may not exceed 0.1 % by weight. This applies to all ships regardless of flag. Regarding reductions in nitrogen oxide emissions from marine engines, Annex VI introduced Tier I, II and III NOx emission standards for new engines. NOx emission limits are set for diesel engines depending on the engine's maximum operating speed. Tier I and II limits are global, while the Tier III standards apply only in NOx Emission Control Areas. Tier II NOx standards are currently being in force. In order to control CO_2 emission from shipping, the first formal CO_2 control regulations were adopted by the IMO, introducing a new chapter, Chapter 4, to Annex VI. Chapter 4 introduces two mandatory mechanisms intended to ensure an energy efficiency standard for ships: the first is the Energy Efficiency Design Index (EEDI), for new ships, and the second the Ship Energy Efficiency Management Plan (SEEMP) for all ships. The regulations apply to all ships of 400 gross tonnage and above and are entered into force on 1 January 2013. Energy Efficiency Design Index (EEDI) is the first globally binding climate change standard. It is anticipated that global CO_2 reductions of 10 to 20 % could be obtained by implementation of EEDI and SEEMP. Detailed descriptions of the emission restrictions prescribed by Annex VI of the MARPOL 73/78 Convention are listed in section 2 entitled 'International regulation concerning air pollution from merchant shipping'. To meet these restrictions on emissions of harmful pollutants from marine diesel engines, different methods and technical solutions can be implemented. Nitrogen oxide reduction methods are generally categorised as primary methods or internal measure and secondary methods or aftertreatment. Primary methods include changes to the combustion process within the engine and can be divided into three main categories: combustion optimisation, water-based control and exhaust gas recirculation, while secondary methods, or aftertreatment, is based on treating the engine exhaust gas itself by passing it through a catalyst system. MARPOL Annex VI will reduce global ship sourced

NOx emissions at a small rate because it only applies to new installations or major conversions. For NOx levels below MARPOL Annex VI, or for retrofitting, the main measures available now are water/fuel emulsions, direct water injection, inlet air humidification and catalyst system. Technical measures to reduce the sulphur oxide emission from ship's diesel engines include the adoption of low-sulphur fuels, the easiest way of reducing sulphur oxides emission. Usage of an exhaust gas cleaning system, i.e. scrubbers, is a possible alternative to low-sulphur fuels to reduce the total emission of SOx and considerably reduces emissions of other polluting particles. A detailed description of these methods is given in section 5 entitled 'Methods of reducing harmful pollutant emissions from marine diesel engines' As the air pollution emissions from ships are continuously growing, it is necessary to constantly improve and implement the efficient technologies and methods in order to reduce pollutant emissions from marine diesel engines and maintain them within the limits prescribed by MARPOL Annex VI as well as by other national and regional regulations. Some of these technologies and methods include the use of shore connection systems of which the main benefits stem from the fact that electricity generated on land by power plants has a smaller adverse impact on the ecosystem than that produced by ship engines, owing to that of particular concern is the pollution generated by ships at berth. Ships equipped with a green technology receive a higher grade. Furthermore, other technical measures for reducing air pollution from ships include the adoption of liquefied natural gas (LNG) as alternative fuel for marine engines. Wind, solar power, biofuels and fuel cells, the world of alternative energy is ongoing and could be useful for reducing air pollution from ships in the future. Finally, harmful pollutant emissions from ships require a stringent international standard due to their impact on the human health and the environment. It includes extending the SOx Emission Control Areas in the EU (e.g. in the Mediterranean, in the Black Sea, in the Irish Sea and in the North East Atlantic) and designating NOx Emission Control Areas as soon as possible.

Author details

Ivan Komar[*] and Branko Lalić

*Address all correspondence to: ivan.komar@pfst.hr

University of Split - Faculty of Maritime Studies - Marine Engineering Department, Split, Croatia

References

[1] Wahlström J, Karvosenoja N, Porvari P. Ship emissions and technical emission reduction potential in the Northern Baltic Sea, Finnish Environment Institute, Helsinki 2006   http://cleantech.cnss.no/wp-content/uploads/2011/06/2006-Wahlstrom-ships-

emissions-technical-emission-reduction-potential-in-northern-Baltic-Sea.pdf   (accessed 17 July 2014)

[2] Friedrich A et al. Air pollution and greenhouse gas emissions from ocean-going ships, The International Council on Clean Transportation (ICCT), 2007 https://www.georgiastrait.org/files/share/PDF/MarineReport_Final_Web.pdf (accessed 17 July 2014)

[3] Air pollution from ships, The European Environmental Bureau, The European Federation for Transport and Environment, Seas At Risk, The Swedish NGO Secretariat on Acid Rain, Updated November 2004 http://www.airclim.org/sites/default/files/documents/shipbriefing_nov04.pdf (accessed 22 July 2014)

[4] Saracoglu H, Deniz C, Kılıç A. An Investigation on the Effects of Ship Sourced Emissions in Izmir Port, Turkey, The Scientific World Journal, Vol. 2013 (2013), Article ID 218324, http://dx.doi.org/10.1155/2013/218324 (accessed 5 August 2014)

[5] Komar I, Lalić B, Dobrota Đ. Air Pollution Prevention from Croude Oil Tankers with Volatile Organic Compound Emission, International Journal of Maritime Science & Technology "Our Sea", Vol. 57, No. 3-4, 138–145, Dubrovnik, 2010. (ISSN: 0469-6255)

[6] Lalić B, Komar I, Nikolić D. Optimization of Ship Propulsion Diesel Engine to Fulfill the New Requirements for Exhaust Emissions, Transactions on Maritime Science (ToMS), April 2014, Vol. 3, No. 1.3 (2014), 1 Split, 2014, 20–31, DOI:10.7225/toms.v03.n01.003

[7] Herdzik J. Emissions from Marine Engines Versus IMO Certification and Requirements of Tier 3, Journal of KONES Powertrain and Transport, Vol. 18, No. 2 (2011), 161–167

[8] Rashidul SMH. Impact of EEDI on ship design and hydrodynamics master of science thesis, Department of Shipping and Marine Technology Division of Sustainable Ship Propulsion, Chalmers University of Technology, Gothenburg, Sweden, 2011, http://publications.lib.chalmers.se/records/fulltext/151284.pdf (accessed 5 August 2014)

[9] Bazari Z, Longva T. Assessment of IMO mandated energy efficiency measures for international shipping, project final report, Lloyd's Register, London, UK, DNV, Oslo, Norway, IMO, 2011, http://www.imo.org/MediaCentre/HotTopics/GHG/Documents/REPORT%20ASSESSMENT%20OF%20IMO%20MANDATED%20ENERGY%20EFFICIENCY%20MEASURES%20FOR%20INTERNATIONAL%20SHIPPING.pdf   (accessed 6 August 2014)

[10] Marty P. Ship energy efficiency study: development and application of an analysis method, PhD Thesis, Universités, École Navale, Nantes, 2014.

[11] Woodyard D. Pounder's Marine Diesel Engines and Gas Turbines, Elsevier, 2004

[12] Komar I, Antonić R, Kulenović Z. Experimental Tuning of Marine Diesel Engine Speed Controller Parameters on Engine Test Bed, Transactions of FAMENA, 33 (2009) 2, ISSN: 1333–1124, Zagreb, 2009, 51–70

[13] ISO 8217:2010/2012 Fuel Standard, Fourth Edition, https://www.bimco.org/News/ 2010/07/19_ISO_8217.aspx (accessed 20 August 2014)

[14] McAllister S, Chen JY, Fernandez-Pello AC. Fundamentals of Combustion Processes, Springer Science+Business Media, 2011

[15] Emission Control Two Stroke Low Speed Diesel Engines, MAN B&W Diesel A/S, 1996   http://www.google.hr/url? sa=t&rct=j&q=&esrc=s&source=web&cd=1&ved=0CBoQFjAA&url=http%3A%2F %2Fwww.flamemarine.com%2Ffiles%2FMANBW.pdf&ei=7nQZVLyUDsXaPPmw-geAI&usg=AFQjCNFBv8IBVUn1vTNfggYmTaJpBLbbjw&bvm=bv.75558745,d.ZWU (accessed 20 July 2014)

[16] Lavoie JB, Heywood JB, Keck JC. Experimental and Theoretical Investigation of Nitric Oxide Formation in Internal Combustion Engines, Combustion Science Technology Vol. 1 (1975), 313–326

[17] Bowman CT. Kinetics of Pollutant Formation and Destruction on Combustion, Progress in Energy and Combustion Science, Vol. 1, No. 1 (1975), 33–45

[18] Kuiken K. Diesel Engines, Part I, Target Global Energy Training; Onnen, 2012

[19] Rakopoulos CD, Giakoumis EG. Diesel Engine Transient Operation - Principles of Operation and Simulation Analysis, Springer-Verlag Ltd., London, UK, 2009.

[20] Technical Bulletin - Combustion Calculations, Normalisations and Conversions, Q Instrument Services Limited, Bishopstown, Cork, Ireland. http:// www.qlimited.com/pdf/Land-Combustion-Calculations-Normalisations-&-Conversions-Q.pdf (accessed 5 June 2014)

[21] Fournier A. Controlling Air Emissions from Marine Vessels: Problems and Opportunities, University of California Santa Barbara, February 2006, http://fiesta.bren.ucsb.edu/~kolstad/temporary/Marine_Emissions__2-11-06_.pdf (accessed 5 June 2014)

[22] Lalić B, Komar I, Antonić R. Water-in fuel emulsion as primary method to reduce NOx from diesel engine exhaust gas, Trends in the Development of Machinery and Associated Technology TMT 2009 Conference Proceedings, 2009, Zenica, Barcelona, Catalunya, Istanbul, 457–460.

[23] Miola A et al. Regulating Air Emissions from Ships, European Commission Joint Research Centre Institute for Environment and Sustainability, Luxembourg, 2010, http://ec.europa.eu/dgs/jrc/downloads/jrc_reference_report_2010_11_ships_emissions.pdf (accessed 5 July 2014)

[24] MAN Diesel and Turbo: Tier III Two-Stroke Technology, https://www.google.hr/?
      gws_rd=ssl#q=MAN±Diesel±and±Turbo%3A±Tier±III±TwoStroke±Technology   (ac-
      cessed 5 August 2014)

[25] Komar I, Antonić R, Matić P. Selective catalytic reduction as a secondary method to
      remove NOx from exhaust gas of the ship's propulsion diesel engine, Proceedings of
      the IFAC Conference on Control Applications in Marine Systems Vukić, Zoran; Sau-
      ro, Longhi (ed.), Zagreb: CEPOST, 2007.

[26] Emission Project Guide MAN B&W Two-Stroke Marine Engines, MAN Diesel & Tur-
      bo, December 2013 http://www.mandieselturbo.com/download/project_guides_tier2/
      printed/7020-0145-001web.pdf (accessed 5 August 2014)

[27] Ship power four - stroke, Wärtsilä Environmental Product Guide, Vaasa, June 2014,
      http://www.wartsila.com/file/Wartsila/en/1278528485383a1267106724867-wartsila-o-
      env-product-guide.pdf (accessed 5 August 2014)

[28]]Lloyd's Register: Understanding Exhaust Gas Treatment Systems - Guidance for
      Shipowners and Operators, London, June 2012, http://www.google.hr/url?
      sa=t&rct=j&q=&esrc=s&source=web&cd=2&ved=0CCQQFjAB&url=http%3A%2F
      %2Fwww.lr.org%2Fen%2F_images%2F213-35826_ECGS-
      guide1212_web_tcm155-240772.pdf&ei=tlMVVNj9F8PXaubw-
      gYAF&usg=AFQjCNGz4a8vCcjsE4e1rsDHx1Rj8Rk0rQ&bvm=bv.75097201,d.bGQ
      (accessed 30 August 2014)

[29] Rr32 Guide to Exhaust Emission Control Options, MS3026, Land & Sea Systems, Bris-
      tol September 99, http://wenku.baidu.com/view/f634f9b81a37f111f1855b5f. (accessed
      10 August 2014)

[30] CNSS: A Review of Present Technological Solution for Clean Shipping, http://
      www.google.hr/url?sa=t&rct=j&q=&esrc=s&source=web&cd=2&ved=0CCIQF-
      jAB&url=http%3A%2F%2Fcnss.no%2Fwp-content%2Fuploads
      %2F2011%2F10%2FSummary-brochure10.pdf&ei=YVkaVPj2KMTfOajQ-
      gYAP&usg=AFQjCNGjfvIe8i4zF_nosPD4Ra1XnSyqqg&bvm=bv.75097201,d.ZWU
      (accessed 5 August 2014)

PERMISSIONS

LIST OF CONTRIBUTORS

B. Maňkovská
Institute of Landscape Ecology, Slovak Academy of Sciences, Bratislava, Slovakia

M. V. Frontasyeva and T. T. Ostrovnaya
Frank Laboratory of Neutron Physics, JINR, Dubna, Russian Federation

Alberto Mendoza and Edson R. Carrillo
School of Engineering and Sciences, Tecnológico de Monterrey, Monterrey, Mexico

Claudia Cappello, Sabrina Maggio, Daniela Pellegrino and Donato Posa
University of Salento, Dept. of Management, Economics, Mathematics and Statistics, Italy

Eliane R. Rodrigues
Instituto de Matemáticas, Universidad Nacional Autónoma de México, Mexico

Mario H. Tarumoto
Faculdade de Ciênicas e Tecnologia, Universidade Estadual Paulista Júlio de Mesquita Filho, Brazil

Guadalupe Tzintzun
Instituto Nacional de Ecología y Cambio Climático, Secretaría de Medio Ambientey Recursos Naturales, Mexico

Thanh-Dong Pham, Byeong-Kyu Lee, Chi-Hyeon Lee and Minh-Viet Nguyen
Department of Civil and Environmental Engineering, University of Ulsan, Daehakro, Namgu, Ulsan, Republic of Korea

Sheikh Saeed Ahmad, Rabail Urooj and Muhammad Nawaz
Department of Environmental Sciences, Fatima Jinnah Women University, Mall Road, Rawalpindi, Pakistan
BZ University, Multan, Pakistan

Kazuo Shimizu
Shizuoka University, Japan

Elaine Patrícia Araújo, Divânia Ferreira da Silva, Shirley Nobrega Cavalcanti, Edcleide Maria Araújo and Marcus Vinicius Lia Fook
Department of Materials Engineering, Federal University of Campina Grande, Brazil

Márbara Vilar de Araujo Almeida
Department of Civil Engineering, Federal University of Campina Grande, Brazil

Mark A. Wolters
Shanghai Center for Mathematical Sciences, Fudan University, Shanghai, China

C.B. Dean
Department of Statistical and Actuarial Sciences, Western University, London, Canada

Ivan Komar and Branko Lalić
University of Split - Faculty of Maritime Studies - Marine Engineering Department, Split, Croatia

Index

www.ingramcontent.com/pod-product-compliance
Lightning Source LLC
Chambersburg PA
CBHW061950190326
41458CB00009B/2834